Ebola Virus Disease

Ebola Virus Disease

From Origin to Outbreak

Adnan I Qureshi

AMSTERDAM • BOSTON • HEIDELBERG • LONDON
NEW YORK • OXFORD • PARIS • SAN DIEGO
SAN FRANCISCO • SINGAPORE • SYDNEY • TOKYO

Academic Press is an imprint of Elsevier

Academic Press is an imprint of Elsevier
125 London Wall, London EC2Y 5AS, UK
525 B Street, Suite 1800, San Diego, CA 92101-4495, USA
50 Hampshire Street, 5th Floor, Cambridge, MA 02139, USA
The Boulevard, Langford Lane, Kidlington, Oxford OX5 1GB, UK

ISBN: 978-0-12-804230-4

British Library Cataloguing-in-Publication Data
A catalogue record for this book is available from the British Library

Library of Congress Cataloging-in-Publication Data
A catalog record for this book is available from the Library of Congress

For information on all Academic Press publications
visit our website at http://store.elsevier.com/

Working together
to grow libraries in
developing countries

www.elsevier.com • www.bookaid.org

Publisher: Sara Tenney
Acquisition Editor: Jill Leonard
Editorial Project Manager: Fenton Coulthurst
Production Project Manager: Karen East and Kirsty Halterman
Designer: Ines Maria Cruz

Typeset by TNQ Books and Journals
www.tnq.co.in

Printed and bound in the United States of America

Contents

10.   Ebola Virus Disease: Psychological and Social Aspects

11.   Treatment of Ebola Virus Disease: Therapeutic Agents

12.   Ebola Virus: Natural Reservoirs

Contributors

Dr. Adnan I. Qureshi, M.D. Zeenat Qureshi Stroke Institute, St Cloud, MN, USA and Zeenat Qureshi Clinical Neurosciences Institute, Conakry, Guinea.

Dr. Omar Saeed, M.D. Clinical Research Fellow, Zeenat Qureshi Stroke Institute, St Cloud, MN, USA

Dr. Morad Chughtai, M.D. Clinical Research Fellow, Zeenat Qureshi Stroke Institute, St Cloud, MN, USA and Zeenat Qureshi Clinical Neurosciences Institute, Conakry, Guinea.

Dr. Nauman Jahangir, M.D. Clinical Research Fellow, Zeenat Qureshi Stroke Institute, St Cloud, MN, USA

Dr. Mohammad R. Afzal, M.D. Clinical Research Fellow, Zeenat Qureshi Stroke Institute, St Cloud, MN, USA

Dr. Ahmad A. Malik, M.D. Clinical Research Fellow, Zeenat Qureshi Stroke Institute, St Cloud, MN, USA

Dr. Mushtaq H. Qureshi, M.D. Clinical Research Fellow, Zeenat Qureshi Stroke Institute, St Cloud, MN, USA

Dr. Ihtesham A. Qureshi, M.D. Clinical Research Fellow, Zeenat Qureshi Stroke Institute, St Cloud, MN, USA and Zeenat Qureshi Clinical Neurosciences Institute, Conakry, Guinea.

Author Biographies

DR ADNAN IQBAL QURESHI

Dr Qureshi is a professor of neurology, neurosurgery, and radiology at University of Minnesota. He has written over 550 scientific publications in prestigious journals including the *New England Journal of Medicine*, *Lancet*, *Archives of Internal Medicine*, *Critical Care Medicine*, *Neurology*, *American Journal of Medicine*, and *Circulation*. In addition, he has made over 1300 presentations in various national and international meetings. He has also been invited as a visiting professor to universities in the United States and abroad. He is the editor of several textbooks and serves on editorial boards for several peer-review journals. He is the principal investigator of a large multinational clinical trial funded by National Institutes of Health. He laid the foundation of the Zeenat Qureshi Institutes in several countries. Since its inauguration, the institutes have led the way in cutting-edge research in epidemiology, clinical trials, and basic research. Most recently, he laid the foundation of Zeenat Qureshi Clinical Neurosciences Institute in Conakry, Guinea. The institute has performed tow studies on Ebola virus disease survivors. The first one has been published in "Clinical Infectious Disease."

DR OMAR SAEED

Dr Omar Saeed graduated from a medical school in Pakistan after which he pursued his interest in the field of medicine focusing on neurology and neuroimaging. He was given the opportunity to work with Dr Adnan Qureshi at the Zeenat Qureshi Stroke Institute where he worked as a clinical research fellow. His main interests were stroke, intracerebral hemorrhage, and neuroimaging among others. During his time at the stroke institute, Dr Saeed has coauthored several scientific publications in prestigious journals including *Journal of Neurosurgery*, *Journal of Cerebrovascular Disease and Stroke*, *Journal of Neuroimaging*, and *Journal of Vascular and Interventional Neurology*. He was also given the opportunity to present at both national and international forums including the International Stroke Conference and The World Masters China Tour and Neuro Interventional International Forum. Furthermore, Dr Saeed has coauthored numerous research abstracts presented at the International Stroke Conference and the American Academy of Neurology

Annual meeting. His future hopes are to continue focusing on both the clinical and research side of medicine. He served as the primary coordinator for the book on Ebola virus disease.

DR MORAD CHUGHTAI

Dr Morad Chughtai graduated *Summa Cum Laude* from American University of Antigua College of Medicine in 2014. As a medical student, he served as a teacher's assistant for multiple subjects throughout the duration of his basic science education, including anatomy, physiology, biochemistry, genetics, pathophysiology, and neurosciences. After obtaining his medical degree, he worked under the instruction of Dr Adnan I. Qureshi as a clinical research fellow for a year's duration. During that time, he coauthored several peer-reviewed publications, one being published in the prestigious journal, *Clinical Infectious Diseases*. He received an award at APPNA annual meeting in 2015 for his work on Ebola virus disease survivors. Subsequently, he obtained a fellowship from the Rubin Institute of Advanced Orthopedics, Sinai hospital in Baltimore, Maryland. He currently serves as an orthopedic research fellow, where he is a coinvestigator of several prospective clinical trials, among other projects. He hopes to carry his passion of teaching and research throughout his medical career and eventually practice in an academic setting.

DR NAUMAN JAHANGIR

Dr Nauman Jahangir completed his medical studies in Pakistan and after completing his postgraduate training in Pakistan, he perused his dream of training in the US as a neurologist. Dr Jahangir had the opportunity to do many clinical rotations at prestigious hospitals. He started at the Zeenat Qureshi Stroke Institute as a clinical research fellow in 2014. During his time at Zeenat Qureshi Stroke Institute, Dr Jahangir worked diligently and was able to publish many peer-review articles in medical journals. He presented multiple medical abstracts at both the International Stroke Conference and the American Academy of Neurology annual meeting where he presented posters and oral presentations. With all these efforts, Dr Jahangir was accepted as a neurology resident at Southern Illinois University where he is continuing his efforts to save lives.

DR MOHAMMAD R. AFZAL

Dr Mohammad Rauf Afzal is working as a clinical research fellow at Zeenat Qureshi Stroke Institute. His research interests include stroke epidemiology, intra-arterial and intravenous thrombolytic use in acute ischemic stroke and neurocritical care. His work has been presented at International Stroke Conference (ISC) and American Academy of Neurology (AAN) annual meeting. He has also coauthored scientific publications in prestigious journals. Along with his research activities, Dr Afzal is also working as an assistant to core image analyst and as an assistant to primary study coordinator in Hennepin County Medical

Center (HCMC) Minneapolis, MN, for Antihypertensive Treatment of Acute Cerebral Hemorrhage-II trial.

DR AHMED A. MALIK

Dr Ahmed Malik is a medical doctor by education and is currently a clinical research fellow at the Zeenat Qureshi Stroke Institute. After completing some undergraduate coursework at SUNY Stony Brook University, Dr Malik, guided by a desire to understand health care in the developing world, went to the Shifa College of Medicine in Islamabad, Pakistan, to complete his medical education. Upon his return to the US, he worked on a volunteer basis with doctors at private clinics before joining Zeenat Qureshi Stroke Institute. As a clinical research fellow at Zeenat Qureshi Stroke Institute, Dr Malik has worked under the mentorship of Dr Adnan I. Qureshi to author numerous scientific research papers published in renowned medical journals. Dr Malik has also presented scientific research papers at national and international medical conferences, is a reviewer for the *World Journal of Pediatrics*, and a contributor to the *Journal of Vascular and Interventional Neurology*. In his free time, he likes to write and is currently writing a few works of fiction for publication.

DR MUSHTAQ H. QURESHI

Dr Mushtaq Qureshi is working as a senior clinical research fellow at Zeenat Qureshi Stroke Institute. He has authored and coauthored several scientific publications which are published in various prestigious journals. He has also made several platform and poster publications in various national and international meetings. Currently, he is the managing editor of *Journal of Vascular and Interventional Neurology* and also serves in the role of the head of imaging department in a large phase III clinical trial, which is funded by National Institutes of Health.

DR IHTESHAM A. QURESHI

Dr Ihtesham A. Qureshi is a clinical research fellow in neurology at Zeenat Qureshi Stroke Institute. His research mainly includes stroke and cerebrovascular diseases, neurointervention, and Ebola virus disease. He has few scientific publications in prestigious journals like Plos One, Elsevier, and F1000 Research. He was also a former medical field doctor at Doctors without Borders/Medecins Sans Frontieres and was involved in operational research on tuberculosis and malaria. He is the principal investigator for the first study ever done to assess memory decline among Ebola virus disease survivors using minimental scale examination. He is also the principal investigator for the first study done to identify persistence of Ebola virus in the synovial fluid among Ebola virus disease survivors using arthrocentesis technique. With the collaboration with Ebola treatment unit, Donka National Hospital, Conakry, Guinea, he was involved in several Ebola virus disease-related research projects.

Acknowledgments

A special thanks has to be extended to Fatoumata Binta Diallo and her family who provided motivation, encouragement, and support for the journey to Guinea. The mission of the journey would not be possible without active collaboration of numerous colleagues in Guinea such as Drs L. Béavogui Kézély and Oscar Loua who opened their doors to us as if we were close friends and family. The book is a tribute to the everyday heroes in West Africa which includes every physician, nurse, and medical professional who evaluates hundreds of sick patients on a daily basis not knowing which will be the next Ebola virus-infected patient that may pass the often fatal disease to them. We pray to God Almighty to give strength and blessings to families who have been affected directly or indirectly from the Ebola virus disease outbreak in their backyards. We also hope that in North America and Europe, the spirit of helping fellow humans prevails over the extreme concerns of self-preservation. Each outbreak is a teacher to humanity and lessons are not limited to medical professionals but are valuable to all that form the class of humanity. We hope that this book provides part of the course material required for broad transmission of such a lesson.

Chapter 1

Ebola Virus: Nature's Experiment Gone Wrong

Chapter Outline

An experiment comprises of dynamic observations made after addition, deletion, or modification of one or more components within the study perimeters. "Nature's experiment" is a term used to define experiments where humans do not plan addition, deletion, or modification processes or the components of study perimeters. Nature's experiments range from the evolution of domesticated dogs from wolf ancestors to the extermination of dinosaurs. In medicine, the term has been used to identify observations of human growth made after genetic defects to the study of etiological agents in disease epidemics. While such experiments are usually the basis of creativity and innovation, the uncontrolled nature of dynamic processes leaves us vulnerable to undesirable consequences. Any experiment where undesirable effects exceed the benefit of the observation is essentially an experiment "gone wrong." Nature's experiments are no exception to such a rule.

The Ebola virus infection may have started as a disease within nonhuman primates, initially termed as "green monkey fever," in order to maintain the fine balance between the number of nonhuman primates and resources available. The competition for survival among nonhuman primates is fierce among the rain forests of Africa as the habitat continues to succumb to the ever-increasing human incursion. Within this "experiment," less-virulent forms of infectious agents are likely to persist because transmission requires diseased primates to survive long enough to contact one another. Most virulent forms of infectious agents may be their own worst enemy by exterminating the very life sources necessary for survival.

But why did the experiment to preserve the African rain forest habitat go so wrong? Why did the Ebola virus start a pandemic in a world already lured to a false sense of security after its victory over smallpox and polio viruses? Why could no one make a vaccine against the virus, centuries after West Africans had developed and practiced smallpox inoculation and well before the colonial rule started using variolation? A West African woman described her childhood experience in 1890s, "they used to scratch your arm until the blood came and then they got the fluid from someone who had the smallpox and rubbed it in"?[1] During the smallpox epidemic of 1954, Director General Sanner of the Public Health for French West Africa stated, "Time is the only means to diminish the ransom being paid by the people" while awaiting for a cure.[1] In contemporary times, the world would also face a rude awakening that it could not stop the growing Ebola virus disease epidemic.

THE FIRST GLOBAL PANDEMIC

Leonardo Giovanni was the only physician within the fleet of ships that sailed out of Kaffa, Crimea into the Caspian Sea, March 1347. The Italians aboard had a lot to celebrate because they were the lucky few who managed to escape the besieged trading post. The Mongols had launched an attack on Italian merchants in Southern Russia in March 1346. The attack culminated in the siege of last trading station in the region, Kaffa (today Feodosia) in Crimea.

The port stayed in Italian hands due to cold winters and mysterious disease decimating the ranks of the Mongol armies. Leonardo had seen numerous funeral proceedings among the besieging armies from the walls of the port town and had attributed these occurrences to the wrath of God and cold weather. But, the winters had been an ideal breeding ground for disease due to close proximity among individuals huddled in close quarters with limited aeration to survive the cold. Leonardo would also be a witness to perhaps the first biological warfare effort with dead corpses infected with disease being catapulted across the walls by the Mongols. His friend Gabriele de' Mussi summarized the events as "Whereupon the Tartars (Mongols), worn out by this pestilential disease, and falling on all sides as if thunderstruck, and seeing that they were perishing hopelessly, ordered the corpses to be placed upon their engines and thrown into the city of Kaffa. Accordingly were the bodies of the dead hurled over the walls, so that the Christians were not able to hide or protect themselves from this danger, although they carried away as many as possible and threw them into the sea."[2]

For now, Leonardo rejoiced with his companions as they approached the port of Constantinople in May 1347. Three of the sailors were sick and one had died during the journey and Leonardo had been responsible for the care of these sailors. After transferring the sick sailors to local infirmaries, Leonardo chose to spend a few days in the city. It had been a long and arduous winter followed by a hectic and rushed journey and Leonardo had numerous reasons to feel exhausted.

By the third day, Leonardo was exhausted and weak beyond reasons and had developed a high fever and chills. He noticed prominent and painful swelling in his groin and axillary regions. He felt utterly unable to walk and his companions took him to the local infirmary. Leonardo recognized the person lying on the next bed as one of the sailors that he had taken care of who now appeared moribund. The sailor's fingers were black-colored and the skin had started to erode from nails onward. Leonardo saw himself as a mirror image of his patient.

By evening, Leonardo slipped in and out of consciousness. He was able to piece together that he had acquired the disease by direct contact with sick sailors. There were others out there who had helped him care for his crew members and they needed to be warned of the impending risk. But Leonardo did not live to warn his companions. He was found dead in the infirmary by the next morning.

Leonardo did not witness the horrifying events that were to follow as the disease spread through his companions and their contacts into Constantinople. By July, ships arriving in Alexandria from Constantinople had spread the disease into North Africa and the Middle East. By the second week of September, ships arriving at the port of Marseilles from Constantinople carrying Leonardo's peers sealed the fate of Europe. The Black Death also known as bubonic plague had come to ravage Europe like no natural calamity or wars to follow. The initial descriptions refer to this plague as the "great pestilence." In a poem composed around 1350, Simon de Covinus described the great pestilence as the mors atra (literally black or terrible death).[3] In the coming centuries, the term "Black Death" will be coined either because of translation of the term mors atra or from the identification of blue–black spots that developed on the skin of infected persons.[3,4] Cities like Pistoia in Italy tried to implement rules and regulations on city inhabitants, closely resembling present principles of quarantine. Travel to any plague-infected area and return was forbidden. No linen or woolen goods were imported into the city and no corpses were buried in the city. However, the city succumbed to Black Death despite strict enforcement of such rules. Today, historians wonder why the disease spread so quickly, a phenomenon that cannot be explained by direct contact or travel routes. The disease overcame natural and man-made fortifications with considerable ease. By the time, plague died down, 25 million people, 60% of all European population, had perished with it.[5–7]

What caused the plague would continue to mystify researchers and philosophers alike in the centuries to come. The disease was caused by the bacteria *Yersinia pestis* transmitted by fleas.[8] The bacteria would be discovered by Alexandre Yersin in Hong Kong in 1894. A member of the French Colonial Health Service investigating the outbreak, he isolated from buboes the bacteria that was later named *Yersinia pestis* after him.

MANKIND VULNERABLE NO MORE?

The next 600 years would see the development and availability of advanced hospital care, highly trained personnel, antimicrobial agents, infection-control strategies, and global coordination efforts turn the world into an impregnable fortress against mass spread of diseases called epidemics. The global eradication of smallpox symbolized the ultimate victory of mankind over disease. The World Health Organization, Resolution WHA33.3 on May 8, 1980, declared, "the world and its peoples have won freedom from smallpox, which was a most devastating disease sweeping in an epidemic form through many countries since the earliest time, leaving death, blindness and disfigurement in its wake and which only a decade ago was rampant in Africa, Asia, and South America."[9–11]

But skeptics wondered whether scientific advancements were enough to keep up with a world that now comprised of population-dense cities cloaked with pollution and a world that has overcome barriers to rapid travel across continents, the ideal breeding ground for disease with no boundaries.

AND THEN CAME EBOLA VIRUS PANDEMIC

It was a quiet Monday morning on March 17, 2014, with the darkness of the night withering into the dusk of dawn. Mahmood was finishing his shift as part of his requirements for completion of his medical school studies. The attending physician asked Mahmood if he could evaluate a patient that had just arrived and was waiting in the emergency room. Mahmood walked into the emergency room like he had done numerous times during his rotations. He found a young woman presenting with fever, agitation, conjunctival hyperemia, dyspnea, and blood clot obstructing the nasal cavity. The examination was based on standard practice and did not require any precautionary measures during direct contact with the patient.

Mahmood reported the findings of his examination to the attending physician and made arrangements for hospital admission. The patient's condition had deteriorated by the time an attending physician made his evaluation. Due to worsening dyspnea, the attending physician placed an oxygen mask and chest electrodes for monitoring electrocardiographic recordings. At 11 am, on March 18, 2014, the patient was dead.

It was the fourth day since Mahmood had evaluated the woman in the emergency room. Mahmood was feeling sick with fever, headache, nausea, and generalized weakness. Mahmood first noticed the fever and fatigue, which was not usual for him given the very active life he lived. He felt too sick to go to work in the hospital. His first thought he had was the most common disease among any medical practitioner living in West Africa: malarial infection. Mahmood made his way to the local pharmacist and purchased antimalarial medication. The initial dose of antimalarial medications, however, did not appear to help him.

Mahmood made his way to the national hospital Ignace Deen and received intravenous fluids. He continued to experience worsening headaches, nausea, and weakness. He felt that the best way was to call the ambulance driver of his hospital to bring him to his hospital for admission. When he arrived at the hospital, tensions were high in the hospital because the attending physician and three nurses who were part of the care team in and after Mahmood's shift had been admitted with similar symptoms. Two of the nurses were close friends and had been caring for the first patient as a team since her admission. Another physician, a neurologist, was admitted the same day with similar symptoms, and traced his contact to evaluation and treatment of a sick child at a private clinic. All of the medical professionals involved were being transferred to Donka National Hospital for isolation and treatment.

On March 23, 2014, the WHO was notified of an outbreak of Ebola virus disease in Guinea.[12] Unfortunately, Mahmood and his medical colleagues were exposed during the time when Ebola virus infection was just emerging. The threat to medical professionals was poorly recognized and medical professionals were caught off guard.

Mahmood was admitted to the isolation unit and was treated with Orasel (oral rehydration solutions), intravenous fluids, metronidazole (for fever and diarrhea caused by parasites), and paracetamol (for pain and fever). Lack of appetite and nausea were quite prohibitive in maintaining adequate nutritional supply. The last days of Mahmood's hospitalization were complicated by delirium. Some of the patients in the isolation unit notified Mahmood that he was verbalizing words that did not make any sense and acting in a nonpurposeful manner. Another patient told Mahmood that he noticed Mahmood wanting to jump over an imaginary fence a few hours prior. Mahmood later recalled other patients who appeared delirious in the third week of hospitalization. The most memorable moment was when a delirious patient started demonstrating aggressiveness and even tried to pull away the mask of one of the physicians trying to feed him.

By the fourth week, Mahmood's ordeal was over. A polymerase chain reaction test demonstrated that he was infection free. He left the isolation unit to go home on April 11, 2014. His mother called him and asked him to come and live with her for a few days. On April 12, Mahmood traveled to the town of N'zérékoré to spend time with his family. Mahmood continued to suffer from severe fatigue and weakness and, at times, delirium. It took weeks of rest before Mahmood was able to return to work. His colleagues were not so lucky. Mahmood would be the only one to survive among the six medical professionals concurrently admitted to Donka National Hospital.

Mahmood would live to see the outbreak spread and become ranked as the largest outbreak of Ebola virus disease by June 18, 2014.[13] On August 8, WHO declared the epidemic to be a "public health emergency of international concern."[12] Mahmood would be one of the approximately 1000 people who would survive the Ebola virus infection among 3343 confirmed Ebola virus disease cases in Guinea, Liberia, Nigeria, and Sierra Leone as of September 14, 2014.[12] Today, Mahmood provides medical care to patients with Ebola virus infection admitted to the isolation unit of Donka National Hospital with unparalleled courage and dedication. Mahmood would become the head of the "Ebola Children Orphans" program in Guinea and the program would strive to provide support to the orphans who have suffered the loss of their parents due to Ebola virus disease.

REFERENCES

1. Schneider WH. Smallpox in Africa during colonial rule. *Med Hist* 2009;**53**. [cited 12/17/2014]. Available from: http://journals.cambridge.org/download.php?file=%2F4263_3226190007041 DC3FB45484B935E6211_journals__MDH_MDH53_02_S002572730000363Xa.pdf&cover= Y&code=340c456c17a9d0992cf84720c13bb8d5.
2. Jones K. *The siege of Kaffa and the black death history in an hour*. 2011. [cited 12/17/2014]. Available from: http://www.historyinanhour.com/tag/black-death/#sthash.05xrQ4En.dpuf.
3. Jones K. *The black death: exposing the myths and lesser-known facts*. 2011. [cited 12/17/2014]. Available from: http://www.historyinanhour.com/2011/01/24/the-black-death-exposing-the-myths-and-lesser-known-facts/.
4. The black death [cited 12/17/2014]. Available from: http://www.theblackdeath.co.uk/.
5. Olea RA, Christakos G. Duration of urban mortality for the 14th-century Black Death epidemic. *Hum Biol* June 2005;**77**(3):291–303. PubMed PMID: 16392633. Epub 2006/01/06. eng.
6. Tsiamis C, Poulakou-Rebelakou E, Tsakris A, Petridou E. Epidemic waves of the Black Death in the Byzantine Empire (1347–1453 AD). *Infez Med* September 2011;**19**(3):194–201. PubMed PMID: 22037442. Epub 2011/11/01. eng.
7. DeWitte SN, Wood JW. Selectivity of black death mortality with respect to preexisting health. *Proc Natl Acad Sci USA* February 5, 2008;**105**(5):1436–41. PubMed PMID: 18227518. Pubmed Central PMCID: PMC2234162. Epub 2008/01/30. eng.
8. Hawgood BJ. Alexandre Yersin (1863-1943): discoverer of the plague bacillus, explorer and agronomist. *J Med Biogr* August 2008;**16**(3):167–72. PubMed PMID: 18653838. Epub 2008/07/26. eng.
9. Langefeld TW, Engel J, Menges T, Hempelmann G. Small pox–infection, therapy and anaesthesiological management (part 1). *Anasthesiol Intensivmed Notfallmed Schmerzther* July 2003;**38**(7):445–55. PubMed PMID: 12822115. Epub 2003/06/25. Pocken–Klinik, Therapie und anasthesiologische Aspekte (Teil 1). ger.
10. Jezek Z. 20 years without smallpox. *Epidemiol Mikrobiol Imunol* August 2000;**49**(3):95–102. PubMed PMID: 11040489. Epub 2000/10/21. Dvacet let bez nestovic. cze.
11. Strassburg MA. The global eradication of smallpox. *Am J Infect Control* May 1982;**10**(2):53–9. PubMed PMID: 7044193. Epub 1982/05/01. eng.
12. Gbandia S. *Ebola spreads to Sierra Leone capital of Freetown as deaths rise: Bloomberg.* 2014. [cited 12/03/2014]. Available from http://www.bloomberg.com/news/2014-07-12/ebola-spreads-to-sierra-leone-capital-of-freetown-as-deaths-rise.html.
13. Dixon MG, Schafer IJ. Ebola viral disease outbreak–West Africa, 2014. *Morb Mortal Wkly Rep* June 27, 2014;**63**(25):548–51. PubMed PMID: 24964881. Epub 2014/06/27. eng.

Chapter 2

A Personal Journey

The enduring legacy of any effort is not what we have achieved, but what we have become as a consequence. Therefore, the shortest effort or journey may not provide the most enduring legacy.

Qureshi AI. Core of Wisdom. AuthorHouse, Bloomington, IN, 2007

ARRIVAL IN CONAKRY, GUINEA

Doctors Adnan I. Qureshi and Morad Chughtai were driving in the car to get to Donka National Hospital, Guinea. They had left the hotel at 9 am after a nice breakfast. It was a warm but humid morning and the sky appeared hazy with clouds in front. The visiting physician had left a chilly autumn climate behind in Minnesota, United States. Both had successfully made the trip from Paris to Conakry the night before on 747 jet from Air France after a quick layover in Mauritania. The crowded airport and traffic jam on the muddy road from airport to the hotel made their final arrival to the hotel even more satisfying. Both had survived the night in Guinea including a mosquito barrage during an outdoor dinner. The intermittently working air conditioner had provided a comfortable night.

As the car passed through the front door of Donka National Hospital (Figure 2.1), both the physicians were able to recognize the entrance, which they had seen in photographs on various Web sites. At this point, adrenaline began to concentrate. The reality of entering the realm of an Ebola virus disease isolation unit was enticing. The anxiousness from arriving at the Donka Airport suddenly seemed like a child's play in comparison to what lay ahead.

Fortunately, Dr Oumar Bailo Niane, one of the leading pharmacists in Guinea, accompanied both of them to the hospital. He was instrumental in guidance for their mission. His presence provided some comfort as he helped them

FIGURE 2.1   Dr Morad Chughtai standing outside the Donka National Hospital.

navigate the Ebola virus-stricken region. It was clear that he had a fear of Ebola virus instilled in him as he offered the foreign doctors a hand sanitizer from his pharmacy several times, explaining to them that the mainstay of Ebola virus infection prevention in Guinea was hand washing.

The group's first stop was office of the administrator of the Donka National Hospital, Dr Hadja Fatou Sikhi Camara. At this point, Dr Camara was not available and had actually left the premises to attend a meeting with World Health Organization (WHO) representatives. From the balcony of the second floor office, the first view of the isolation unit could be visualized. Smoke was going into the air above the unit as Ebola virus-riddled equipment was being burned. Dr Ahmed Malik, one of the team members from St. Cloud, Minnesota, had already established contact through e-mail with Dr Kezley, the chair of neurosurgery at the Donka National Hospital. Dr Malik was scheduled to accompany Drs Qureshi and Chughtai to Guinea but his visa did not get approved in time. Dr Kezley was a dynamic and highly qualified person. He had done neurosurgery training in Moscow, Russia, and yet had returned back to Guinea, to provide service to his people. Dr Kezley had already outlined the plan to meet the key leaders, involved nationally in management of the Ebola virus disease epidemic.

Our first stop was Dr Moumier Barry (Figure 2.2), the head of infectious diseases at Donka National Hospital. He had the overall responsibility of supervising the response to the Ebola virus outbreak at Donka National Hospital and any affiliate facilities that transferred patients to Donka National Hospital. He had a friendly and welcoming personality and they had not received many visitors from the United States. Their main liaison had been with Institute Pasteur in Paris, France, and Doctors Without Borders teams that had been in Donka National Hospital for the last 6 months.

FIGURE 2.2   Outside department of infectious and parasitic diseases, Donka National Hospital, Conakry, Guinea. Left to right: Dr Kezley Beavogui, Dr Morad Chughtai, Dr Moumier Barry, Dr Adnan I. Qureshi, and Dr Oumar Bailo Niane.

Communication, however, appeared to be an issue because neither Dr Qureshi nor Dr Chughtai spoke French and Dr Malik, the colleague with fluency in French, was not able to come on the trip. The communication transpired with broken English and none-the-less was quite productive. Dr Barry patiently answered all of the visitors' questions and thanked the visiting physicians for their efforts to step forward and help in the control and treatment of Ebola virus infection in Guinea. He was also kind enough to explain the national response in terms of identified patients or suspected patients of Ebola virus infection and the pathway in terms of notification, ascertainment, and transfer of the patients to the Donka National Hospital facility.

The Donka National Hospital was the main hospital of two hospitals designated for the treatment of patients with Ebola virus infection. Dr Barry provided a detailed overview of the current setup in Guinea. The first station was at Donka National Hospital, Conakry, and the second one was in Gueckedou for Ebola virus disease patients located far away from Conakry. The Donka National Hospital had an 85-bed treatment facility (Figure 2.3) and is primarily managed by Doctors Without Borders. Conakry has a population close to 2 million[1] and is thought to contain a quarter of the entire country's population. Gueckedou is an area in Guinea where it is bordered by Sierra Leone (Kenema) and Liberia (Foya) and also houses an 85-bed facility. The population in this area is close to 250,000.[1] Both of these centers have established laboratories.

A transit center in Macenta, east of Gueckedou, is a 35-bed center. The patients admitted to this facility are eventually transferred to Conakry or Gueckedou. This transit center has no laboratory and a 30-bed treatment unit in Macenta

FIGURE 2.3   View of Ebola virus disease isolation unit (high-risk zone).

is currently in progress, but should be ready by the end of 2014, and hence may serve as a third-field station. Although there are no current laboratories being built for this center, Dr Barry explained the current challenges and also blessed any efforts to promote health care for patients with Ebola virus infection.

Soon after Dr Kezley was able to contact a trainee of his, Dr Oscar Loua, who was a neurosurgery resident and fortunately fluent in French and English. He had self-taught himself English, and now also performed duties for the United States Embassy providing translations for meetings and documents. His addition to the group enhanced the quality of communication immensely. He had no clinical responsibilities during the outbreak because his mentor, Dr Kezley, did not want him to enter the hospital to protect him from acquiring Ebola virus infection. Although Dr Loua's courage tremendously increased the productivity of the group, his break from clinical responsibility made it clear that Ebola virus disease continued to ravage Guinea.

The next stop was a meeting with Dr Sidiki Diakité, the head of the national Ebola virus infection response in Guinea and the personal liaison to the president of Guinea. Dr Kezley pulled in front of the hospital with his car and Drs Chughtai, Qureshi, and Loua got in. The whole group drove through the busy streets of Guinea and finally made it to the office of the headquarters for the Ebola virus infection response in Guinea. The building was small, made up of bricks with patches of different colors. The group was quickly led to the office of Dr Diakité (Figure 2.4).

Dr Diakité had been kind enough to spend time with the team although he had an impending meeting with the president of Guinea to update him on the situation in the country. Dr Diakité was kind enough to listen to the objectives of the visit and provided blessing for their efforts. The group spoke about developing protocols for pain and agitation control. Dr Qureshi explained some of his

FIGURE 2.4   Meeting at Ministry of Health, Conakry, Guinea. Left to right: Dr Kezley Beavogui, Dr Adnan I. Qureshi, Dr Sidiki Diakité, and Dr Morad Chughtai.

experience in developing such protocols based on his training background in neurocritical care in the United States. Dr Diakité personally thanked the visitors for the visit and answered the specific questions in terms of a bigger picture response in Guinea. He also informed the team about the recently approved trial vaccine against Ebola virus infection, sponsored by the WHO. The meeting was quick and concluded with a group photograph to serve as memorabilia. It was time to go back to the hotel and Dr Kezley was kind enough to drop the visiting physicians, Drs Chughtai and Qureshi, to their respective destination hotel. The tentative plan was laid out for the next day.

TOUR TO EBOLA VIRUS DISEASE ISOLATION UNITS

The next day was initially cloudy with intermittent spurts of rain; however, as Dr Chughtai and Dr Qureshi finished breakfast and moved toward the car, the weather appeared to have cleared up. The hotel was a premium resort for visitors coming to Guinea for vacation and pleasure. Unfortunately, because of the Ebola virus infection scare, the hotel appeared empty (Figure 2.5). The breakfast hall that was usually crowded in the morning hosted only Dr Qureshi and Dr Chughtai.

The car ride to the Donka National Hospital took around 15 min, and upon arrival, Dr Loua was already in front of Dr Kezley's office on the fourth floor of the hospital. Dr Kezley appeared shortly after and all members of the group went into his office to exchange greetings. A few minutes later, Dr Elhadj Ibrahima Bah joined the meeting (Figure 2.6). He was the man in charge of care of Ebola virus disease patients admitted at the Donka National Hospital. Additionally, he acted as the main liaison between the Donka National Hospital and Doctors

FIGURE 2.5 A once full dining hall at the Mariador Palace Hotel now empty because of the Ebola virus disease outbreak and travel warning to endemic areas.

FIGURE 2.6 A meeting with Dr Bah to discuss Ebola virus disease research. Left to right: Dr Adnan I. Qureshi, Dr Kezley Beavogui, Dr Elhadj Ibrahima Bah, and Dr Tokpagnan Oscar Loua.

Without Borders, an organization that had overall command of the medical care in the Donka National Hospital isolation unit. Dr Bah was an energetic man, but obviously under a lot of pressure. He took his mission very seriously and was willing to take risks that came with the job. It was clear that he was facing pressures from being the main liaison to the Doctors Without Borders personnel and from his family and relatives who did not accept him working in the Donka National Hospital Ebola virus disease isolation unit.

FIGURE 2.7   Inside Ebola virus disease isolation unit (low-risk zone), Donka National Hospital, Conakry, Guinea. Left to right: Dr Adnan I. Qureshi, Dr Morad Chughtai, and Dr Tokpagnan Oscar Loua.

Dr Bah accompanied the visitors to the whole Ebola virus disease isolation unit set in the center of Donka National Hospital. The group entered through an entrance point following the mandatory hand wash with 0.05% chlorine solution and the shoes were sprayed with 0.5% chlorine solution. The group proceeded inward over a muddy track impregnated with bricks and gravel. The central track zigzagged through the center of the isolation unit and numerous health workers glanced at the visiting team. There was a customary exchange of greetings with the physicians of Doctors Without Borders. Dr Bah explained the risk and social stigma that each of these health workers faced when they worked at the Donka National Hospital Ebola virus disease facility. These risks included being shunned from their families and even hostile behavior from their relatives and fellow villagers. Therefore, great effort was made not to identify the healthcare workers and to avoid photographs that may reveal identifying features.

Dr Bah led the visitors to an enclave where various components of the isolation unit (Figure 2.7) could be clearly visualized. The acute care of Ebola virus disease patients was conducted in this isolation ward, which comprised of a single-level building structure, an open area delineated by temporary fencing, and an encampment in front of the building serving as the entry and exit points. On the right was the outdoor encampment consisting of several tents which housed suspected Ebola virus disease patients and those who had recent contact with Ebola virus disease patients. Each person was observed there for a few days and was required to have three polymerase chain reaction (PCR) tests for detection of Ebola virus infection. If all the three tests did not detect any evidence of infection, the person was free to leave the encampment. A young man was departing from the encampment after being declared free of Ebola virus infection and the signs of relief were obvious on his face as he passed the visiting physicians. If these patients were confirmed with Ebola virus infection, they would have move to the closed building in front of the enclave.

All confirmed Ebola virus disease patients had beds within the building. Patients were located and shifted based on several criteria but mainly availability. Protective gear up prior to the entry and gear down upon exit from the building was undertaken at the encampment. The open area had a section for Ebola virus disease which confirmed patients to come out and meet visitors. Two Ebola virus disease patients, a middle-aged woman and young man, could be seen sitting in the visiting area conversing with family members whose backs were toward the visitors. The visiting physicians looked at the two patients from a distance and were pleasantly surprised not to see any obvious signs of the disease in any of them. The visitor seating was separated by 10 feet temporary fences, and there was an adjacent open location where the fully geared member of the treating team came out and loudly read the notes regarding each patient including vitals, medication administered, and any new developments. The notes were being dictated by two health-care personnel sitting 10 feet away.

Dr Bah went over the day-to-day working within the isolation unit. All patients within the isolation unit had been confirmed as having the Ebola virus infection by means of a PCR test. The PCR test is done at two places in Guinea, Gueckedou and Donka Hospital, and the results are available after 2 h. The number of PCRs performed vary on a daily basis depending upon the number of patients admitted that day. There were 46 Ebola virus-infected patients in the acute care unit at the time of visit, which included two medical students, two physicians, and two nurses. Acute care was provided in three shifts starting from 7am to 2pm, followed by 2–8pm, and 8pm–7am. There were five teams within each shift, which rotated across the isolation unit in a sequential manner resulting in one team within the unit at a given time. The team members were doctors, nurses, psychologists, and hygienists. Each team member were protective gear, including headgear, gloves, goggles, boots, and apron. The team members washed hands at several locations prior to entry with chlorine solution (0.05%). The protective gear was decontaminated by spray of chlorine solution (0.50%) upon exiting the isolation unit. Any item that entered the isolation unit is considered contaminated and cannot be brought back out. The patients' excreta and clothes were burned (Figure 2.8).

The patients' symptoms were broadly grouped into three categories: fever and headaches, vomiting and diarrhea, and bleeding which can occur from any part of the body. There were occasional symptoms of confusion, restlessness, and agitation. Blood pressure and pulse rate were measured four times daily using cuff measurements and manual palpation within the unit, respectively. There were no mechanical ventilator available and supplemental oxygen via face mask was the only option for respiratory distress. There were no antiviral medication, vaccine, or exchange transfusion from survivors in use. Blood transfusions were very rarely used and no intravenous medications (which were

FIGURE 2.8   View of the Ebola virus disease isolation unit showing sandbags surrounding fire pit where contaminated disposable materials are burned daily.

routinely available in intensive care units in the United States) were available to treat low blood pressure. Laboratory tests such as complete blood count and serum electrolyte measurements were not performed because samples had to be sent out to routine laboratories in other parts of the hospital where the risk of contamination and infection spread cannot be contained.

Each patient was treated with antimalarial agents, which was a combination of artemether and lumefantrine (Figure 2.9). The treatment focused on the symptom relief, which included oral acetaminophen for fever, headache, and severe joint and muscle pains. Tramadol was also used for pain control and intravenous or intramuscular morphine sulfate was infrequently used when other medication failed. Oral hydration formulation was the preferred route for maintaining adequate hydration. Intravenous hydration using intravenous fluids rich in dextrose (glucose) or electrolytes was used to treat dehydration, low blood pressure, and malnutrition. Nutritional care for Ebola virus-infected patients was based on food provided by families, inspected at the Ebola treatment center, and subsequently administered to patients. Patients who were not able to tolerate oral intake rely on caloric intake from intravenous fluids rich in glucose. There were no other nutrition formulations in use that could be administered intravenously.

The average stay within the isolation unit was 2 weeks. If a patient was without any fever, afebrile, or asymptomatic for 3 days, a PCR test was performed and, if negative for the presence of virus, the patient was discharged. The survival of confirmed Ebola-infected patients was 60% with most survivors being discharged home. The discharge symptoms included fatigue, anorexia, and persistent joint pains. Infrequently, patients were transferred to other units in the hospital after being infection free because of other acquired infections such as

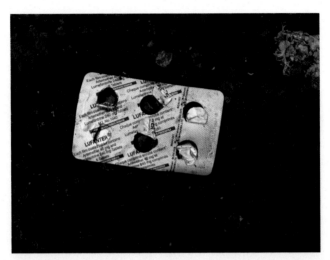

FIGURE 2.9 A dose pack of artemether and lumfantrine (drugs used to treat malaria) lying on the streets of Conakry.

pneumonia. The most common cause of death in Ebola virus-infected patients was severe dehydration and low blood pressure, a condition commonly referred to as hemodynamic shock.

Owing to the busy schedule of Dr Bah, a second meeting was planned in the afternoon. The visitors walked past the most visible landmark within the encampment, a large tree with unclear origin. The whole encampment was over-shadowed by a large tree with a trunk that was several feet thick and the lush green foliage provided a soothing and calm environment within the encamp-ment. If the tree had eyes, it would have been in tears having witnessed death and tragedy within the encampment on a daily basis.

As the group exited through the encampment, some of the health-care workers requested a portion of the hand sanitizer from the bottle hanging like a necklace from Dr Chughtai's neck. A 50-cent bottle of hand sanitizer in the United States available from any retail store was in demand in Guinea. Dr Chughtai carefully dispensed doses of the hand sanitizer to those who requested for it. It was evident in his dispensation technique that he was trying to avoid direct skin contact with anyone. He may have been generously dispensing, but was meticulously avoiding potential contact with individuals with possible contact with infected individuals.

The second meeting was again held in the encampment in the central enclave, which was within the encampment but not an isolation area. Dr Bah introduced the group to a young woman: the third patient admitted to the Donka National Hospital at the beginning of the epidemic. She was one of the seven people in her family who had contracted the Ebola virus infection and one of the three who had actually survived the infection. Today, she volun-teered her time to provide care to the patients who were infected with Ebola virus infection, serving as a spokesperson for hope. Her survival was a proof to each patient admitted to the isolation wards that their outcome was not a death sentence. Survival and return to normal life were still a possibility. Dr Bah went over the patient charts and explained the charting process, which included vital signs and any new development in clinical symptoms or signs. The cool night within the encampment was enlightened by electrical bulbs and the mosquitoes flew around in dozens. The fires within encampment were visible and almost mesmerizing as the undesirable products were incinerated in large holes dug within and around the encampment. It was almost like a bonfire celebrating another day of the epidemic moving behind them, despite its ongoing nature. There was a clear reduction in the chaotic buzz that the isolation unit usually gave off during the day.

A VISIT TO THE VILLAGES

Our second to last day in Guinea proved to be as exciting as the previous ones. Several reports indicated that the incidence and even prevalence of Ebola virus disease were higher in the forest regions. Whether such higher prevalence was a result of unique eating habits of certain tribes and consumption of what is called as bushmeat or essentially just a consequence of routine close contact between individuals in these villages remained unclear. The group wanted to visit the villages to gain better understanding of the potential inciting causes of higher incidences of Ebola virus infection in the forest. However, the villages that were affected the most were around 800 km away from Conakry and a single day of travel was not possible. These villages did not have airports to accommodate short flights and the ground transportation appeared to be only practical way to get to these villages. There were villages that were closer to Conakry and the group decided to visit those villages (Figure 2.10) to gain better understanding of the conditions that exist there.

The road trip leaving the capital was an exciting one. The driver navigated skillfully through roads that were partially formed, sometimes of mud and sometimes of concrete and most of the time a mixture of both. The paths that acted as roads were uneven; sometimes it was as if the car was tilted 45° to one side. It was easy to see that the people who complain about driving in high traffic region such as New York City and New Jersey perhaps need a drive on the roads in and outside Conakry to fully appreciate their blessings. Despite the difficult driving conditions, the scenes of Guinea outside the city were spellbinding. The lush, green rain forest was both enchanting, and enormously addictive.

The group drove on for over an hour and finally exited the road to enter the village. The drive through mud inroads was slow and painstaking. The village appeared to be a combination of mud brick buildings and straw huts. Central gathering of individuals within the village was a common routine. It was

FIGURE 2.10   A visit to a village in Guinea.

not uncommon to witness hostility of the individuals in the villages because of widespread anger due to media exploitation during the current epidemic. As the group proceeded through the village, villagers strongly opposed using the camera for taking the photographs. Dr Loua's gentle interference was able to calm matters with the villagers. The group proceeded farther into the bush forest and the road conditions got worse. The driver skillfully navigated through the thick foliage and uneven roads without showing any exasperation or stress.

After 45 minutes drive into the hard mud roads, the group came to a point where all the other cars turned back. The drivers said that a body had been found two miles from here and the authorities sealed off the area due to concerns that the person had died under uncertain circumstances. The first thought in everybody's mind including the visiting physicians from the United States was it might be an Ebola virus infection victim who had died in the village. While scientific curiosity prompted the visiting physicians to go farther, the desire was overshadowed by concerns of not violating the isolation policy of the existing law enforcement officials and an investigation that was already occurring in the village.

The group turned around and went into a different direction until a point was reached where the river was flowing rapidly enriched by continuing rainfall (Figure 2.11). Drs Loua, Qureshi, and Chughtai walked across the river on stepping-stones. Members of the group were able to see some of the local practices of using these rivers for bathing and other water-based necessities. Of most interest were the tourist groups from the Middle East who were having a barbecue along the banks of the river. Another young person was sitting under a tree away from the main crowd smoking what appeared to be

FIGURE 2.11   Drs Qureshi and Loua pose for a photograph on river bank.

Marijuana. He appeared very calm and at peace with his surroundings. He stared into the river as if it was his only getaway from the daunting pressure of Ebola virus disease devastating his nation. The Ebola virus disease epidemic had obviously not dampened the enthusiasm for life among tourists and the local population.

TIME TO BID FAREWELL

The last day in Guinea was uneventful but sad. The electricity was in and out throughout the day and Drs Qureshi, Chughtai, and Loua spent the afternoon in the shades of the balcony of the hotel. There was an unescapable lure to the people of Guinea and the work that remained to be done within the Donka National Hospital. We planned continued collaboration with the doctors there.

Dr Niane came to say goodbye to the hotel. That was a tough day for him because one of his relatives had died from Ebola virus disease at the Donka National Hospital. He was in the midst of arranging the funeral proceedings to be carried out in the mosque in front of the Donka National Hospital. His relative had contracted Ebola virus disease in Sierra Leon but had returned to his village in Guinea despite the blockade to road travel between countries. One of Dr Niane's uncles had notified him and Dr Niane had arranged for local authorities to meet and transport the relative to the hospital. Unfortunately, the isolation had not happened early enough and the wife and mother of the diseased relative had also contracted the disease and were admitted to the Donka National Hospital. We gave him our condolences and felt sad that we were leaving without visiting them.

The drive to the Conakry Airport was short and the airport was less crowded than the day of the arrival. Dr Loua accompanied Drs Qureshi and Chughtai and bid farewell at the doors of exit control. A last mandatory chlorine solution hand wash was necessary to enter through exit control and there were two screening posts within the airport to screen every person departing from Guinea for the presence of fever. A cutaneous swap was performed and both Drs Qureshi and Chughtai were declared free of fever. Dr Chughtai was given an additional full body thermal scan that was being tested experimentally.

Inside the departure area, an unexpected meeting took place with two members from the United States Centers for Disease Control and Prevention (CDC) who were visiting the airport to ensure that methods were in place to contain the disease within travel zones. Both were members of a three-member team from the CDC, placed in Guinea for a one-month tour. Both were originally from Atlanta, like many other members of the CDC, and Dr Qureshi shared some of his experiences when he was a visiting scientist in the organization almost two decades ago. None of them had visited a hospital or had any direct contact with Ebola virus disease patients, and there were two other teams in Sierra Leon and Liberia. The last stop prior to boarding the plane required another temperature check and fortunately none of the passengers were detained due to fever.

RETURN HOME AND THE COLD WELCOME

The return back to the United States was not pleasant. Dr Qureshi returned to Minneapolis St. Paul International Airport where his name was on the entry control list due to his travel to Guinea. After full declaration, the Transportation Security Administration and passport control personnel provided a printed set of instructions to identify symptoms of Ebola virus disease and to report to the hospital in vicinity in the event of any symptoms. In the absence of symptoms, quarantine was not considered necessary. Dr Chughtai entered from Dulles International Airport and had a similar set of instructions provided to him.

Dr Qureshi experienced many more hostilities in the next few days. His academic employer, the University of Minnesota, reported to the Minnesota Department of Health that he had human brains in his possession, a rumor spread to discredit his efforts. In desperation, the director of Employee Health Clinic requested Dr Qureshi to not come to the University of Minnesota premises for at least 21 days. The director acknowledged that there were no medical grounds for such a request, but was merely for the psychological satisfaction of employees at the University of Minnesota. At St. Cloud Hospital, the vice president of medical affairs called a meeting and openly criticized Dr Qureshi, questioning his judgment.

Despite Dr Qureshi demonstrating his compliance with the existing federal and state regulations, he was now a pariah within the hospital. His hopes of sending unused medical equipment to the Donka National Hospital in Guinea were a thing of the past. No one was interested in learning about his experiences and what each of them as an individual or as a group could do to help people in need. It appeared that funding efforts seem to be misguided, more-so focused on preventing Ebola virus disease from reaching the United States, rather than directly improving the mortality rate in Africa. The level of airport screening seemed unnecessary and had not proved to detect travel-associated Ebola virus-infected cases. The screenings were done to comfort the general United States population.

The money going into funding airport "comfort measures" could have saved many lives. Research shows that electrolyte and other laboratory findings in the blood can effectively guide precision of treatment for dehydration and malnutrition. A monitoring setup based on automated blood pressure, electrocardiographic monitoring, and electronic displays within the isolation unit and secondary distant station can allow early detection of hypotension and institution of corrective measures. Availability of point-of-care laboratory testing for hemoglobin, coagulation parameters, and electrolytes within the isolation unit can guide treatment including blood product transfusions and electrolyte replacements. Availability of blood products for transfusion with appropriate processing and storage capabilities within the unit and provision for intravenous medication administration and mechanical ventilator support can reduce fatalities. But these measures were not available in isolation units of West Africa and additional funding could have made a difference.

FINAL THOUGHTS

The United States is blinded to the social effects of Ebola virus disease. There are children with no surviving family members who become pariahs in their communities, orphaned by the stigma of being around someone who has already passed from the disease; issues as such should be prioritized over comforting unaffected nations. Many in the United States were caught in a frenzy of self-preservation, unable to distinguish between fear and actuality in regards to their risk of contracting Ebola virus disease. To such medical professionals, despite taking the Hippocratic oath, the ravages of the Ebola virus disease epidemic in Africa thousands of miles away from the United States meant nothing more than a reason to protect their own lives.

Undeterred, Drs Qureshi and Chughtai continued to remain in communication with their comrades in Guinea and plan to return within the next few months with medical equipment and needed supplies. In the subsequent months, they would work closely with Drs Kezley and Loua to lay the foundations of Zeenat Qureshi Clinical Neurosciences Institute in Donka National Hospital, Conakry (Figure 2.12).

FIGURE 2.12 A picture to celebrate the start of Zeenat Qureshi Clinical Neurosciences Institute, Donka National Hospital. Drs Kezley and Chughtai with neurosurgery staff.

REFERENCES

1. populationmondiale.com 2014; Available from: http://www.populationmondiale.com/wg.php? x=&men=gcis&lng=en&des=wg&srt=npan&col=abcdefghinoq&msz=1500&geo=-89#sth ash.Ds35ZVAQ.dpbs.

Chapter 3

Ebola Virus: The Origins

Chapter Outline

After all it really is all of humanity that is under threat during a pandemic.
Margaret Chan, Director-General of the World Health Organization (WHO)

THE HISTORY OF EBOLA VIRUS DISEASE OUTBREAKS

How exactly do we define a pandemic? By most accounts, it is an epidemic of an infectious disease that has spread through human contact to multiple countries or continents. The World Health Organization (WHO) first developed pandemic phases in 1999, namely for the influenza virus, and these phases were later revised in 2005. According to the WHO, there are six phases that occur during a pandemic and three subsequent phases called recovery phases.

Phases 1, 2, and 3 are combined into one called the Uncertain probability of a pandemic. Phase 1 is further described as a period of time when there is no animal to human transmission. Phase 2 occurs when an animal, either domesticated or wild, has shown transmission to a human. Phase 3 occurs when an animal or a human-animal virus has caused a small cluster of disease in people but disease has not yet been transmitted from human to human. Phase 4, which is the Medium to High probability of a pandemic, is described as human to human transmission of an animal or human virus, which has caused a sustained community-level outbreak and has been verified. Phase 5, the High to Certain probability of pandemic, occurs when the same virus has caused a continuous community-level outbreak in at least two countries in one WHO region. Finally, Phase 6 is called pandemic in progress and is when the same virus has caused a continuous community-level outbreak in at least one other country in a second WHO region. The last three phases which could be described as recovery phases are named by WHO as the Post-Peak period, Possible New Wave, and Post-Pandemic Period.[1]

So how does Ebola virus disease, which was previously thought to be an epidemic strictly present in Equatorial Africa, now come under the definition of a pandemic (Figure 3.1)? The growing concerns, fears, and recent news about newly diagnosed patients in the United States and other countries have spread fear among many residents of these countries. The use of the word "pandemic" gives a face to people's fears and provides hope that most humanity will unite together to fight against the disease. But where did it all start and were there ever other cases reported outside of Africa in the past?

"Ebola virus," is a household term, but few recognize it as a filamentous virus that has a negative sense ribonucleic acid genome. The term "virus" originates from a Latin word simply meaning "slimy fluid." Virions are cylindrical or tubular structure containing a viral envelope, matrix, and nucleocapsid component, and are approximately 80 nm in diameter and 800–1000 nm in length. Over a century, this definition has evolved from the first identified human virus, which caused yellow fever reported in 1901 by the US Army physician Walter Reed. The work followed after pioneering work in Cuba by Carlos Finlay, proving that

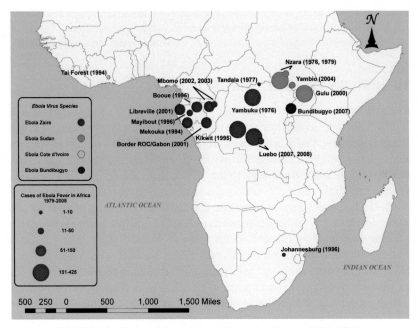

FIGURE 3.1 Geographic locations of Ebola virus disease (1976–2008).

mosquitoes transmitted the deadly disease. The Chamberland–Pasteur filter had been developed in 1884 in Paris by Charles Chamberland, who worked with Louis Pasteur. But it was Dmitri Ivanowski in St. Petersburg in Russia in 1892 who used porcelain filters to isolate and characterize what we now know to be a virus. The name Ebola is derived from the name of the Ebola River near a town called Yambuku in the Democratic Republic of Congo (previously Zaire) where the first outbreak was identified.[2] There have been five subtypes identified, namely, Ebola-Zaire, Ebola-Sudan, Ebola-Bundibugyo, and Ebola-Taï Forest. The fifth which was called Ebola-Reston never actually infected any humans.[2]

MARBURG VIRUS DISEASE EMERGENCE (1967–1975)

The Ebola virus has characteristics that are very similar to another virus from filoviridae family of viruses called the Marburg virus. This virus was first described in summer of 1967 when an outbreak of unknown disease occurred in Germany and Yugoslavia. It was also a filamentous RNA virus source of which was traced to the vervet monkeys (Cercopithecus aethiops) imported from Uganda.[3]

There was another case of Marburg virus diagnosed in a young man who was admitted to a hospital in South Africa in 1975. About a year later, a hemorrhagic outbreak of unknown origin began in the areas bordering Uganda (from origins of the previously mentioned 1967 outbreak in Germany), namely, Zaire and Sudan. The patients presented identical symptoms to those seen in patients with the Marburg virus infection. Samples of the patients' blood were sent to laboratories in England, Belgium, and the United States to identify the responsible pathogenic agent. Although a virus with the same morphology as the Marburg virus was extracted, the viral proteins such as glycoprotein, which are structural components of virus, were different.[4]

EBOLA VIRUS DISEASE IN SUDAN AND ZAIRE (1976–1977)

It remains unclear where the first actual case of Ebola virus disease occurred because the first actual outbreak occurred simultaneously in Sudan and Zaire (now the Democratic Republic of Congo) in 1976. The first identifiable case of Ebola virus disease was reported in Sudan in 1976. In total, out of 24 reported outbreaks of Ebola, 3 have occurred in Sudan during the years 1976, 1979, and 2004 according to the WHO (Figure 3.2).

While the first case of Ebola virus disease was reported in Southern Sudan, the country was in the midst of a hot summer with an average daily high temperature of 102–105 °F. Sudan was in political turmoil in July of 1976 due to the failed 1976 Mercenary Invasion by Al-Sadiq al-Mahdi and Ansars. Sudan's strain of the Ebola virus was first described in 1977.[5] The disease was then described as the Ebola hemorrhagic fever in 1978 by the WHO, which detailed the virus type in the 1976 outbreak.[3] After the 2002 outbreak, the term was later changed and called the Sudan Ebola virus. The virus was designated as the Sudan virus as its final name in 2010.[6]

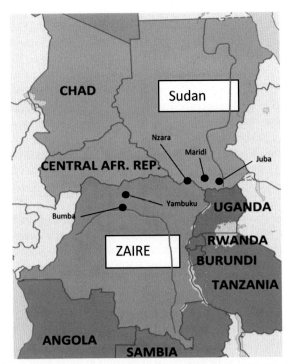

FIGURE 3.2 Map of the first-ever Ebola virus disease outbreak in towns of Sudan and Zaire (now the Democratic Republic of Congo).

The first patient was an employee of a cotton factory in Nzara township who became ill on June 27, 1976.[3] Five days later, he died, and by July 6th, a second case had also died. Two other people from the same cotton factory also developed the disease and eventually passed away due to hemorrhagic manifestation of the disease.[3] The three people who died from the cotton factory were cared for by loved ones who eventually also got the disease and died. One of the workers from the cotton factory who died had a brother who cared for him and also subsequently felt ill and had traveled to Maridi a small town in South Sudan where he was admitted to the Maridi Hospital on August 7, 1976, but he eventually died.[3] This eventually triggered the spread of the disease to neighboring villages. The difference between the two outbreaks was that the Nzara outbreak was spread from workers in the cotton factory to their families and the Maridi outbreak was spread through the hospital setting and health workers.[3]

In total, there were 284 cases: 67 in Nzara, 213 in Maridi, 3 in Tembura, and 1 in Juba, which were neighboring villages. One hundred and fifty-one of these patients died as a direct result of the infection characterized as a "mystery epidemic" in Sudan and news began to appear on the pages of the "Guardian" from October 1, 1976.

A week later, health officials at the Heathrow Airport had been alerted to check on travelers from Sudan and Zaire. WHO admitted that it was having difficulty in identifying the virus, although on the 14th of October it claimed that the virus in question was the Marburg virus disease. Strict isolation procedures were set up by WHO in collaboration with the Sudanese government and sufficient supplies of protective gear were donated by the Center for Disease Control and Prevention (CDC) in Atlanta, Microbiological Research Establishment in Porton, England, and the Government of the Federal Republic of Germany. The last recorded infection in Nzara was on October 27, 1976. By the end of the month, it was reported that the mysterious green monkey disease had disappeared.

The first known case of Ebola virus disease in Zaire was a 44-year-old school instructor who was admitted to Yambuku Mission Hospital on August 26, 1976, with fevers presumed to be malaria at that time.[7] He was suspected of having malaria and treated with a parenteral injection of chloroquine. By September, nine other cases occurred in at the same hospital, suggesting that infection was active within the Yambuku Mission Hospital.[7] The outbreak was traced back to contaminated needles at a hospital, where only five syringes were used each day to treat all the patients. In 1976, a group of researchers headed by Belgian scientist Peter Piot traveled to the remote village of Yambuku in the Republic of Zaire to study what they thought to be the Marburg virus—a hemorrhagic fever virus that was discovered a few years earlier.

The laboratory in Antwerp discovered Ebola virus disease after a blood sample from a Belgian nun who had fallen mysteriously ill in Zaire was given to Peter Piot. On October 13, 1976, Frederick A. Murphy, a virologist for CDC and an expert in photographing viruses took the first photograph of the virus. A few days prior to the incident, a blood specimen from the infected patient in Zaire was delivered to the CDC. At that time the culprit was thought to be the Marburg virus. While in transport, the tubes with blood specimens had been shattered, but a virologist by name of Patricia Webb managed to salvage some of the remaining blood, which was subsequently mixed with monkey kidney cells. Days later, she noticed the monkey cells looked damaged so she gave Murphy the samples to visualize under a microscope. At the time, Murphy was the only expert at the CDC who had looked at another virus called Marburg. Only three people had been allowed to study it because it was so dangerous, and two had left the agency. Murphy called Webb and Karl Johnson, head of CDC Special Pathogens. After a couple of hours and additional tests, Webb confirmed that the virus was not Marburg, but something entirely new.

Two groups worked simultaneously on this newly emerging virus which had caused devastation in both Zaire and Sudan in 1976. The first group focused more on the isolation of the virus and the effects of inoculation of the virus on both adult and newborn mice. Blood samples were taken from a 42-year-old woman who was infected in Yambuku, Zaire. Serum from the blood was then inoculated into six adults and two litters of newborn mice both intracerebrally (meaning in the brain) and into vero cell cultures (which are cell lineages used in cell culture). The virus was found to be very lethal in that it killed two adult mice on the fourth and fifth day; after which brain biopsy samples were taken from both dead and surviving mice. In the newborn litters, only one survived from first litter and second litter had three left including one dead, one paralyzed and one very sick. Samples were also taken from these mice. In vero cells, cytopathic (damage of host cell by virus) effects were seen by the 12th day. Electron microscopy showed what might have been the very first visualization of the Ebola virus and its effects.[8]

The second group, which was performing research during the same time, had isolated this new unknown virus and observed it under an electron microscope. Visualizations showed that this virus was about 199 nm in diameter and varied between 300 nm and more than 1500 nm. It had surface projections of about 10-nm long and internal cross striations which showed a helical core or spiral-like structure after staining. The structure was very similar to the Marburg virus. Laboratory technicians observed an interesting difference in the Zaire strain: it had more branching of its filamentous particles and also a more continuous envelope which extended beyond its rigid internal structure. Blood and postmortem liver samples were taken from patients in Zaire.[9] This new virus was then named Ebola virus similar in morphology to Marburg virus, but antigenically (process of changing its surface protein in order to evade host immune response) different.

The Health Ministry of Zaire announced in 1976 that the virus responsible for the recent epidemic of green monkey fever that claimed several hundred lives will be known as the Ebola virus, named after a river in the north. A total of 318 cases of confirmed Ebola virus disease were identified between September 1, 1976 and October 24, 1976, out of which 280 died from the disease resulting in a case fatality rate of 88%, one of the highest for any epidemic in recent years. Since the Yambuku Mission Hospital was one of the only major hospitals in Zaire responsible for managing Ebola virus disease patients at that time, the inability to contain the disease within the hospital led to repeated exposures in patients who were inflicted with other diseases and more likely to die due to multiple concurrent illnesses.[7] To date, out of 24 recorded outbreaks of Ebola virus, 9 have occurred in the Democratic Republic of Congo (Zaire) during 1976, 1977, 1995, 2001–2002, 2003 (January–April), 2003 (November–December), 2005 2007, 2008, and 2012 according to the WHO.[10]

A year later in 1977, there was an isolated case of Ebola hemorrhagic fever that was discovered in Tandala, Zaire. It was discovered in a 9-year-old girl who presented hemorrhagic fever-like symptoms resembling Ebola virus disease. The child was from a village by the name of Bonduni and she was taken to the Tandala Mission Hospital located in the northwestern area of Zaire's Congo basin. The girl eventually died from the disease and postmortem specimens were taken to isolate the virus. The interesting part of this isolated case was that the girl had never traveled to any endemic areas of previous outbreaks nor had any family member or village member been sick with a similar disease. Even after her death no family members became ill.[11] Although the first outbreak might have been recorded in 1976, the evidence of Ebola hemorrhagic fever previously existing in the area of Democratic Republic of Congo has been suggested.

When the 1977 isolated case occurred in the Tandala hospital, there was a group of missionaries who were working in the northwestern areas of Zaire, which included hospital staff members from Tandala. Serums were collected from them and tested for the Ebola virus. One of those specimens from a physician came back positive for Ebola virus disease. It was later learned that the doctor had performed an autopsy in May of 1972 on a student from the Bible School who died of a hemorrhagic fever-like illness which was thought to be yellow fever.[11]

EBOLA VIRUS DISEASE OUTBREAK IN NONHUMAN PRIMATES IN COTE D'IVOIRE (1994)

In 1994, there was a troop of chimpanzees who had contracted the virus and demonstrated hemorrhagic manifestations and eventually died in the Tai National Park located in Cote D'Ivoire (Figure 3.3). A Swiss ethnologist who was one of three people performing necropsy on chimpanzees, contracted the Ebola virus disease. In 1994, a new strain of the Ebola virus was isolated from a Swiss researcher with a dengue-like syndrome, who had likely been infected during the necropsy of a chimpanzee.[12] The animal had been found dead in the Taï National Forest (Côte d'Ivoire) during a 2-year epidemic that killed half of the population of chimpanzees. The patient recovered after being treated and, despite lack of containment measures and quarantine, there were no transmission of the disease. Since there was such a high number of fatality in these chimpanzees, they were not thought of as being reservoirs of the disease.[12]

EBOLA VIRUS DISEASE IN GABON (1994–1996)

Between 1994 and 1996, Gabon located in the west coast of Central Africa experienced three different outbreaks which had fatality rates of 60%, 68%, and 75%.[10]

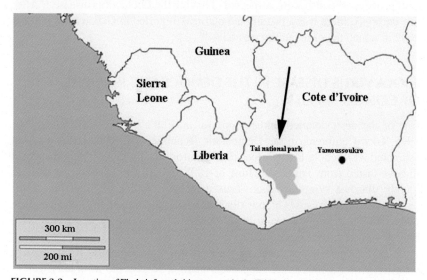

FIGURE 3.3   Location of Ebola infected chimpanzees in the Tai National Park, Cote D'Ivoire. (Arrow)

The Gabon outbreak started in 1994 when there were unexplained deaths reported from gold-mining sites. A total of 32 sick people from the three forest encampments were identified, 23 of which were from Mekouka, 4 from Andock, and 5 from Minkebe. They subsequently traveled to a nearby hospital at Makokou. While widely speculated that a local population of primates (i.e., chimpanzees and gorillas) also contracted the virus and had died, the dead primate bodies were never recovered or found. There were additional cases that occurred after these primary cases, but they were mostly due to the lack of strict containment and quarantine of patients who had arrived from the gold-mining camps. The virus killed 45 out of the 60 cases detected in the capital Libreville and Booue in 1994.

There were two additional epidemics in Gabon in the spring and fall of 1996. In February of 1996, 18 people became ill in the village of Mayibout, Gabon which is 40 km south of Mekauka and Andock where the first outbreak reported previously occurred. The affected had symptoms suggesting Ebola virus infection (fever, headache, and bloody diarrhea); 4 out of 18 died 48 h after the symptom onset. These people had allegedly skinned and chopped up chimpanzee cadavers. In October of 1996, there were samples of blood from patients in a hospital in Booue that tested positive for the Ebola virus. A retrospective investigation showed that this epidemic likely started as early as July with the death of a hunter in a logging camp near Booue. In the next 2 months, two other hunters were also infected. One of these escaped from medical authorities at Booue and died in the village of Balimba where he was being treated by a nganga (spiritual healer) who also contracted the disease and subsequently died.

The epidemic ended in March of 1997 with total of 60 cases and 45 deaths. These were unrelated to the previous gold mine epidemic, but were interestingly correlated to primates due to their involvement with hunting or skinning chimpanzees who also were suspected of having the Ebola virus disease.[13] After this outbreak, there was an additional outbreak reported in Gabon from 2001 to 2002 in which the case fatality was 82%.[10]

EBOLA VIRUS DISEASE IN THE DEMOCRATIC REPUBLIC OF CONGO (1995)

One of the most detailed studies done for an Ebola outbreak was during the 1995 Kikwit outbreak in the Democratic Republic of Congo (DRC) with an estimated 317 people being infected, 245 unfortunately died. It was suspected to have lasted from January to June of 1995. Initially it was thought that the bloody diarrhea symptoms were caused by an outbreak of *Shigella* species (a bacterium which causes bloody diarrhea). After processing 97 blood samples, out of which 23 were actually blood stained upon microscopic examination, only 4 were positive for *Shigella* species.

Experts suspected that the cause was a nosocomial (originating from hospital) infection with the index case supposedly being a 36-year-old laboratory technician who underwent laparotomy at Kikwit General Hospital for suspected bowel perforation. He underwent two laparotomies after which he ended up dying on April 14, 1995. Medical personnel involved in the surgical procedure also became ill, most likely due to contact with blood products of the patient, which led them to believe the infection might be viral in origin. This diagnosis was confirmed when blood from 14 acutely ill patients was sent to CDC by the Institute of Tropical Medicine. Out of those 14 samples, 1 had Ebola virus antigen, 2 had Ebola-specific IgM antibodies, and 12 had Ebola virus RNA, which was detected by using reverse transcriptase polymerase chain reaction (which is a technique used in molecular biology to detect genetic data such as RNA expression).

Control measures were then initiated to limit disease spread. On May 4, 1995, all hospitals, laboratories, and health centers were closed throughout Kikwit; all patients who were in contact with suspected Ebola virus disease patients were isolated and quarantined in a separate building. Some patients and hospital staff fled the hospital due to the fear of being quarantined. The Zairian government declared that an epidemic outbreak had occurred and requested help from the WHO. Protective equipment was given to all health-care workers who were to take care of Ebola virus infected patients. All health-care workers were also trained in the proper use of protective clothing when dealing with potential Ebola virus disease cases. Even burials of patients who had succumbed to the disease were taken care of by trained Red Cross volunteers who wore protective gloves and clothing and disposed of all the bodies in plastic bags to make sure traditional rituals were not used, which often required cleaning of the body by family members and kissing and touching the body, both of which would have spread the disease further.[14]

Patient care was mostly symptomatic before the diagnosis of Ebola virus infection in the 1995 Kikwit outbreak. Patients were thought to have malaria and given antimalarial agents and those suspected gastrointestinal infections were given multiple antibiotics, all of this done by the health personnel without proper protective gear. This lack of preventive measures was one of the main reasons for initial spread of the disease. After the epidemic was recognized, health-care personnel took charge of control measures.

Except for the emergency department, which continued its normal functions, all other activities were focused on controlling Ebola virus disease spread. Isolation wards were cleaned by calcium hypochlorite 2% solution, mattresses were covered with plastic, and bodies of dead patients were disinfected and disposed of in plastic bags. Patients were only treated symptomatically by giving them paracetamol for pain and fever, metoclopramide and haloperidol for nausea, vomiting, and hiccups, and diazepam in case of agitation and epilepsy symptoms. This was also the first time an algorithm for Ebola virus disease patients was developed for an emergency department. Patients with obvious Ebola virus disease symptoms were admitted to an isolation unit. Patients suspected of having Ebola virus disease remained in emergency department for 24 h under observation for development of further symptoms and then moved to an isolation unit.

Conjunctival redness was also apparently highly predictive of the Ebola virus infection, which was noticed by nurses and other health-care personnel. Since eye redness was not routinely seen in diseases such as malaria and gastroenteritis, patients who presented such eye redness were also isolated. Patients with hemorrhagic manifestations were transferred to isolation units. It was difficult to differentiate from other clinical diseases due to the lack of rapid diagnostic tests for Ebola virus infection in patients.[15]

During the last week of the epidemic, there were about 25 patients with the disease. Control measures had been well organized by that time and adequate protective equipment was also readily available. Nurses and physicians took daily detailed notes. Dehydrated patients were given intravenous fluids using disposable butterfly needles, a process which had previously been more contagious. All patients were also better nourished by providing protein-rich foods. The last Ebola virus disease patient was transferred out of the isolation unit and discharged from the Kikwit General Hospital on July 31, 1995.[15]

During this whole ordeal there were about 55 patients out of the total 317 in whom exposure source was not known, and remained unknown after an extensive epidemiologic evaluation. All cases were identified using various techniques: death registry, retrospective, prospective view of clinic and hospital records, door-to-door case findings, and rumor. Patients were interviewed using a questionnaire, which categorized cases into those with or without reported sources of exposure. A probable case of Ebola hemorrhagic fever was defined as having an unexplained hemorrhage or fever and previous contact with another Ebola-infected person, or fever accompanied by any three of these symptoms headache, nausea, vomiting, anorexia, intense fatigue, abdominal pain, myalgia, arthralgia, dysphagia, dyspnea, or hiccups. Confirmed cases were further tested for Ebola antigens or antibodies using an enzyme-linked immunosorbent assay (ELISA—a test which uses antibodies and color change to identify a substance) or the detection of viral RNA using a reverse transcription polymerase chain reaction.

Only those patients who were identified with an unknown source of exposure were included in the study. If the patient was found to be deceased, a household member or close relative was interviewed. These proxies were identified up using a hierarchical system. The first proxy chosen was from the same house as the deceased knew the patient well, next from a different household who knew the patient, and the last chosen were those residing in the same area (i.e., landlord or close neighbor). Results of the study showed that 55 of the 316 patients had no reported source of exposure at the beginning of the investigation. Of these, four patients had survived the illness and were able to participate in the survey. Proxies for the remaining 40 patients provided information for them.

A likely source of infection was identified in 32 (73%) of 44 patients. Seven had visited someone sick with Ebola virus disease symptoms, nine had been admitted to a health-care center in 3 weeks preceding onset of Ebola virus disease symptoms, and six had both risk factors. Out of the 23 who had visited an ill friend or relative with Ebola, 4 resided in the same house, 14 reported touching the ill patient, and 5 visited without touching the patient. The 12 patients who did not have any kind of contact or exposure were used to study the role of social customs, occupational activities, and dietary habits in the acquisition of Ebola virus infection. Of the 23 patients who had previous exposure to a case of Ebola virus disease, 19 had only visited the patient and were not involved in care. None of these according to them had any contact with blood, feces, vomitus, urine, or saliva, but 14 did touch the patient. It remains unknown from this study how the remaining five patients who became infected acquired the disease.[16]

The final aspect of the Kikwit outbreak was to ascertain the potential natural history of the disease and/or natural reservoirs. This was done when background investigation isolated the index case to be a probable 42-year-old man who had come in contact with the natural reservoir of the disease, before previously mentioned disease outbreak was made known to the outside world in April 1995.[13] Although the natural reservoir of this virus was suggested to be in bats, there was no evidence against the transmission through arthropods in one study.[17] Species of arthropods that were collected included mosquitoes, bedbugs, ticks, sandflies, fleas, nonbiting flies, lice, and tsetse flies. Results of the study after collecting 34,985 arthropod specimens, out of which 27,849 were selected for virus isolation. No Ebola virus was isolated from any of the species that were collected.[17]

This outbreak managed to attract unprecedented publicity, which led to an even larger interest by researchers and scientists to study the disease and outbreak. Outcomes of this disease outbreak included of better management, control, and patient care, but the mystery of the disease origin still remains unexplained. Control measures may have led to the end of disease spread, but how and when it started still remain unclear and mysterious.

EBOLA VIRUS DISEASE IN UGANDA (2000–2001)

On October 8, 2000, the Ministry of Health in Uganda received a report from a remote village Rwot-Obillo 14 km north of the Gulu town district about patients who had manifested symptoms that were very similar to hemorrhagic fever. Teams were dispatched immediately to investigate the outbreak and after collection of specimens they were sent to the National institute of Virology in South Africa. The results confirmed the outbreak to be due to Ebola-Sudan serotype.[18]

In October of 2000, the morbidity and mortality weekly report by the CDC published an article about an unusual febrile illness that had started to occur in the Uganda population. The outbreak was described as having hemorrhage and fevers with a very high mortality rate. The cases were reported to the Ministry of Health in Kampala by the superintendent of St. Mary's Hospital in Lacor, and the District Director of Health Services in the Gulu District. There were additional cases found in the Gulu Hospital, which was the regional hospital in that area. Ebola hemorrhagic fever was confirmed when specimens taken from infected patients were sent to the National Institute of Virology in Johannesburg, South Africa.[19]

Control measures were initiated by setting up surveillance teams and clinical case management teams geared toward stopping the spread of the disease. A separate Ebola hemorrhagic fever surveillance system was set up to assess the magnitude of the actual outbreak. Many of the prevention strategies were to follow up with confirmed cases for a period of 21 days, educate about avoidance of traditional burial methods, establish trained burial personnel teams, prevent large public gatherings, and set up isolation wards. It was later confirmed that virus was the Ebola-Sudan strain.[19]

By January of 2001, there were approximately 425 presumptive cases and 224 deaths occurred with a 53% mortality rate. The Ebola virus disease was recorded in 3 districts in Uganda: 393 patients from Gulu, 27 from Masindi, and 5 from Mbarara. The major symptoms reported in 62 of the Ebola virus infected patients were diarrhea, asthenia, anorexia, headache, nausea and vomiting, abdominal pain, and chest pain. Bleeding occurred in about 12 (20%) patients.[19] Of all the recorded Ebola virus disease outbreaks, this was by far the largest. On February 27, 2001, Uganda was declared officially to be free of Ebola hemorrhagic fever, following a 42-day period, twice the maximum incubation period, during which no new cases had been reported.

One of the reasons for rapid and effective control of this outbreak compared with all the others was the rapid response that was initiated by the Ugandan government and early involvement of WHO with setting up field offices and laboratories at outbreak sites. After this outbreak, Uganda had four smaller outbreaks compared with the initial one in 2000. The other outbreaks occurred in 2007, 2011, and two outbreaks in 2012 with attack rates of 25%, 100%, 71%, and 57% respectively (Figure 3.4).[10]

SPORADIC OCCURRENCES OF THE EBOLA VIRUS DISEASE

Over the years, there have been isolated cases of Ebola virus disease. On November 5, 1976, an investigator at the Microbiological Research Establishment, Porton, England, accidently inoculated himself with needle being used to take samples from guinea pigs inoculated with the serum from patients in Sudan outbreak.[3,4] There patient experienced symptoms of the disease but survived.

CASE FATALITY DIFFERENCES BETWEEN EBOLA VIRUS DISEASE EPIDEMICS

Why is there such a difference in mortality among some of the Ebola virus strains? The difference between the Marburg and Ebola viruses as previously mentioned was not morphological but serological. A study done in 1979 identified another difference between the two viruses that the Marburg virus had an outer coat which was more resistant to erosion by negative staining while the Ebola virus was not.[18] Specific strains of Ebola virus were then suspected due to the differences in mortality between strains. It was noticed that the mortality rate of the Zaire strain initially was a more higher than that of the Sudan strain. Some might have attributed that to better control of the disease, but difference at a molecular level could also be plausible.

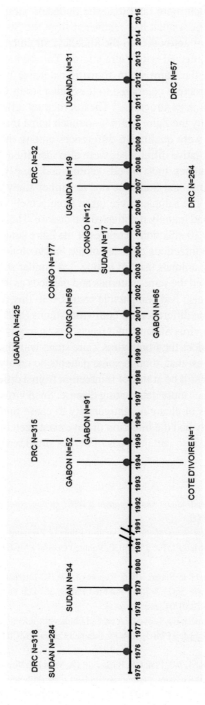

FIGURE 3.4 Timeline of major Ebola virus disease outbreaks over the years: DRC, Democratic Republic of Congo; ●, year of outbreak; N, number of cases.

A comparative study published in 1980 interestingly showed that the Zaire strain proved to be much more fatal than the Sudanese strain in both humans and animals. Even monkeys that had recovered from the Sudanese strains and developed high levels of antibodies to the Sudanese strain became viremic and eventually died when infected with Zaire strain.[5] This lack of immunity could have also been due to the Zaire strain showing extensive branched form and long filaments in its structure compared to the Sudan strain which showed less branching and irregular in its structure.[18] These structural differences could give us some insight into why the Zaire strain is so much lethal than the Sudan strain of Ebola virus. There were qualitative differences among the strains of Ebola virus but a more quantitative differences were also identified.

The quantitative studies took blood, serum, and biopsy specimens from patients of both the Zaire and Sudan strains to further study the differences in its infectivity process. The samples were taken from acutely ill patients in both Zaire and Sudan who had initially manifested the disease. The results of the study showed that fewer than 10 infectious particles of the Zaire strain were lethal compared to nearly 10,000 infectious particles of the Sudan strain, which even then failed to kill any of the animals used in testing.[20] A similar study was also done in which researchers used the same materials and methods as the previously mentioned study and they also had comparable results.[21] In this study, they showed that there were antigenic differences in both virus strains (Sudan and Zaire). The antiserum of the Ebola virus Sudan strain reacts less aggressively with the Ebola virus Zaire strain than does the Ebola virus Zaire strain with the Sudan strain.[21]

These studies suggest that, though some patients do recover from the Ebola virus disease they may still be at risk of reinfection from a different strain that is more virulent and causes more debilitating disease. Such viruses do exist which have similar properties of changing antigenicity to cause reinfection in people who might have already had the infection or have even been immunized such as the influenza virus and also certain types of herpes simplex virus types.

REFERENCES

1. *Pandemic influenza preparedness and response: a WHO guidance document*. Geneva: World Health Organization; 2009.
2. Zhang L, Wang H. Forty years of the war against Ebola. *J Zhejiang Univ Sci B* September 2014;**15**(9):761–5. PubMed PMID: 25183030. Pubmed Central PMCID: PMC4162877. Epub 2014/09/04. eng.
3. Ebola haemorrhagic fever in Sudan, 1976. Report of a WHO/International study team. *Bull World Health Organ* 1978;**56**(2):247–70. PubMed PMID: 307455. Pubmed Central PMCID: PMC2395561. Epub 1978/01/01. eng.
4. Emond RT, Evans B, Bowen ET, Lloyd G. A case of Ebola virus infection. *Br Med J* August 27, 1977;**2**(6086):541–4. PubMed PMID: 890413. Pubmed Central PMCID: PMC1631428. Epub 1977/08/27. eng.
5. Bowen ET, Lloyd G, Harris WJ, Platt GS, Baskerville A, Vella EE. Viral haemorrhagic fever in southern Sudan and northern Zaire. Preliminary studies on the aetiological agent. *Lancet* March 12, 1977;**1**(8011):571–3. PubMed PMID: 65662. Epub 1977/03/12. eng.

6. Kuhn JH, Becker S, Ebihara H, Geisbert TW, Johnson KM, Kawaoka Y, et al. Proposal for a revised taxonomy of the family Filoviridae: classification, names of taxa and viruses, and virus abbreviations. *Arch Virol* December 2010;**155**(12):2083–103. PubMed PMID: 21046175. Pubmed Central PMCID: PMC3074192. Epub 2010/11/04. eng.

7. Ebola haemorrhagic fever in Zaire, 1976. *Bull World Health Organ* 1978;**56**(2):271–93. PubMed PMID: 307456. Pubmed Central PMCID: PMC2395567. Epub 1978/01/01. eng.

8. Pattyn S, van der Groen G, Courteille G, Jacob W, Piot P. Isolation of Marburg-like virus from a case of haemorrhagic fever in Zaire. *Lancet* March 12, 1977;**1**(8011):573–4. PubMed PMID: 65663. Epub 1977/03/12. eng.

9. Johnson KM, Lange JV, Webb PA, Murphy FA. Isolation and partial characterisation of a new virus causing acute haemorrhagic fever in Zaire. *Lancet* March 12, 1977;**1**(8011):569–71. PubMed PMID: 65661. Epub 1977/03/12. eng.

10. Organization WHO. Table: Chronology of previous Ebola virus disease outbreaks [cited 12.02.14]. Available from: http://www.who.int/mediacentre/factsheets/fs103/en/.

11. Heymann DL, Weisfeld JS, Webb PA, Johnson KM, Cairns T, Berquist H. Ebola hemorrhagic fever: Tandala, Zaire, 1977–1978. *J Infect Dis* September 1980;**142**(3):372–6. PubMed PMID: 7441008. Epub 1980/09/01. eng.

12. Formenty P, Boesch C, Wyers M, Steiner C, Donati F, Dind F, et al. Ebola virus outbreak among wild chimpanzees living in a rain forest of Cote d'Ivoire. *J Infect Dis* February 1999;**179**(Suppl. 1): S120–6. PubMed PMID: 9988175. Epub 1999/02/13. eng.

13. Georges AJ, Leroy EM, Renaut AA, Benissan CT, Nabias RJ, Ngoc MT, et al. Ebola hemorrhagic fever outbreaks in Gabon, 1994–1997: epidemiologic and health control issues. *J Infect Dis* February 1999;**179**(Suppl. 1):S65–75. PubMed PMID: 9988167. Epub 1999/02/13. eng.

14. Muyembe-Tamfum JJ, Kipasa M, Kiyungu C, Colebunders R. Ebola outbreak in Kikwit, Democratic Republic of the Congo: discovery and control measures. *J Infect Dis* February 1999;**179**(Suppl. 1):S259–62. PubMed PMID: 9988192. Epub 1999/02/13. eng.

15. Guimard Y, Bwaka MA, Colebunders R, Calain P, Massamba M, De Roo A, et al. Organization of patient care during the Ebola hemorrhagic fever epidemic in Kikwit, Democratic Republic of the Congo, 1995. *J Infect Dis* February 1999;**179**(Suppl. 1):S268–73. PubMed PMID: 9988194. Epub 1999/02/13. eng.

16. Roels TH, Bloom AS, Buffington J, Muhungu GL, Mac Kenzie WR, Khan AS, et al. Ebola hemorrhagic fever, Kikwit, Democratic Republic of the Congo, 1995: risk factors for patients without a reported exposure. *J Infect Dis* February 1999;**179**(Suppl. 1):S92–7. PubMed PMID: 9988170. Epub 1999/02/13. eng.

17. Reiter P, Turell M, Coleman R, Miller B, Maupin G, Liz J, et al. Field investigations of an outbreak of Ebola hemorrhagic fever, Kikwit, Democratic Republic of the Congo, 1995: arthropod studies. *J Infect Dis* February 1999;**179**(Suppl. 1):S148–54. PubMed PMID: 9988178. Epub 1999/02/13. eng.

18. Okware SI, Omaswa FG, Zaramba S, Opio A, Lutwama JJ, Kamugisha J, et al. An outbreak of Ebola in Uganda. *Trop Med Int Health* December 2002;**7**(12):1068–75. PubMed PMID: 12460399. Epub 2002/12/04. eng.

19. Outbreak of Ebola hemorrhagic fever Uganda, August 2000–January 2001. *Morb Mortal Wkly Rep* February 9, 2001;**50**(5):73–7. PubMed PMID: 11686289. Epub 2001/11/01. eng.

20. McCormick JB, Bauer SP, Elliott LH, Webb PA, Johnson KM. Biologic differences between strains of Ebola virus from Zaire and Sudan. *J Infect Dis* February 1983;**147**(2):264–7. PubMed PMID: 6827142. Epub 1983/02/01. eng.

21. Richman DD, Cleveland PH, McCormick JB, Johnson KM. Antigenic analysis of strains of Ebola virus: identification of two Ebola virus serotypes. *J Infect Dis* February 1983;**147**(2): 268–71. PubMed PMID: 6827143. Epub 1983/02/01. eng.

Chapter 4

Ebola Virus Disease Epidemic in Light of Other Epidemics

Chapter Outline

All epidemics are based on an outbreak of disease caused by virus or bacteria, which might or might not be new to humans, and spread through various methods, ranging from contaminated food to person-to-person transmission.

39

COMMON FEATURES OF EPIDEMICS

According to previous research,[1,2] several factors can start an epidemic including the following:

1. Disasters (e.g., wars, famine, floods, and earthquakes)
2. Temporary population settlements
3. Preexisting diseases in the population
4. Ecological changes like floods and cyclones
5. Resistance potential of the host (i.e., nutritional and immunization status of the host)
6. Damage to public utility and interruption of public health services

There are three patterns of disease continuity in epidemics. The "sawtooth" pattern is where there are intermittent outbreaks of a disease that recede in intensity, but the disease is not eradicated from the population. The smallpox epidemics in Africa during the 1920s–1950s would be an example of such a pattern. The "tooth necklace" pattern is where the disease is eradicated from the population, but pathogen species is kept alive under controlled circumstances for preparation of vaccines and biological studies. While the escape of pathogen from confinements of laboratories has been the subject of numerous conspiracy theories, vaccination with live attenuated viruses is more likely to be the string to maintain the continuity.

The third pattern is the "tooth eruption" pattern where, like the tooth hidden within the gums and emerging independent of other teeth, the pathogen emerges and is exterminated without any relation to previous occurrences. The Ebola virus is one of the pathogens following the "tooth eruption" pattern where the disease emerged in Sudan and the Democratic Republic of Congo (DRC) in 1976, disappeared, and then reemerged in Uganda between September 2000 and February 2001, only to mysteriously disappear. It emerged again in December 2013 in Guinea.[3]

What is far more perplexing is why epidemics die their deaths, a phenomenon noticed since the beginning of humanity. While it is convenient to believe that measures such as vaccination of at-risk individuals, quarantine of diseased persons, and acute and timely treatment are the cause of epidemic eradication, the facts do not support such a conclusion. In fact, the largest epidemics, such as the Peloponnesian War Pestilence, Antonine Plague, Plague of Justinian, Black Death of the fourteenth century, and Spanish flu, came to an end without widespread use of any of those strategies.

There are three main theories for the spontaneous remission of epidemics:

1. There are two types of people within the exposed population: some more vulnerable and some more resistant. The people who may be resistant to the disease may be so because of previous exposure to viruses with similar structures resulting in the development of immune responses that are adequate for multiple pathogens. They might also be resistant due to superior health, including age and nutritional, and occupational advantages. The virus might eventually be faced with a population that is completely resistant to the infection.

2. Changing environment within habitats that are not conducive to the survival or propagation of viruses or other pathogens. Weather changes, including temperature and humidity fluctuations, may significantly influence the survival or propagation of a virus outside the body. Elimination of reservoirs that carry pathogens including animals, insects, food, or water, by chance or design, may disrupt the cycle of propagation. Such elimination of infection is less likely to occur within an epidemic because of diverse factors and geographical areas involved.

3. The most likely explanation is the "Sand Filter Theory," a term coined by Adnan I. Qureshi, MD. This theory reflects the similarity between retention of particulate matter during filtration based on density of sand particles, which can be compared to pathogens within a population based on population density. Most epidemics are composed of diseases that require close contact between diseased and healthy individuals for continued propagation of pathogens. Unlike natural disasters, such as hurricanes, floods, volcanoes, and changes in climate that exist independent of population density, epidemics depend upon population density, a feature shared with reproduction rates, migrations, and predation. After population density reduces below a critical limit, such contact may not be available enough for continued propagation of pathogens.

SPANISH FLU EPIDEMIC (1918–1920)

One of the most monumental of twentieth-century epidemics, the "Spanish flu" influenza pandemic in 1918, infected 25–30% of world's population and resulted in death of almost 40 million people.[4] The world had seen flu epidemics before. The influenza epidemic occurred in Europe in 1580s, started in Russia and spread to Continental Europe and Africa followed by another epidemic in 1743.[5] A more devastating flu epidemic happened in 1830–1833. The term "Spanish flu" was a misnomer because the disease did not originate in Spain. The disease was rampant in Germany, Britain, France, and the United States; wartime censors minimized early reports of illness and mortality in these countries.[6] During the 1918 flu pandemic, Spain's king, Alfonso XIII (1886–1931), became very ill. His illness and recovery from the disease was reported to the world because Spain was neutral and was not under wartime censorship restrictions, while outbreaks of flu in other belligerent countries were concealed. This created the wrong impression that Spain was most affected and caused the pandemic dubbed as the "Spanish flu."[7] Even President Woodrow Wilson (1856–1924) reportedly contracted the flu in early 1919 while negotiating the Treaty of Versailles, which ended World War I.[8]

The epidemic appeared in two phases. The first one appeared in late spring of 1918, known as the "3-day fever," without any warning and resulted in few deaths and victims recovered after a few days. The typical symptoms of the flu were chills, fever, and fatigue resulting in a low number of deaths. During that same year in the fall, however, a highly contagious and deadly wave of influenza emerged. Victims died within hours or days of symptom onset. Their skin turned blue, and patients' lungs filled with fluids. The flu did not discriminate between rural and urban areas ranging from densely populated East Coast to sparsely populated parts of Alaska. Young adults were among the hardest hit group, a group that usually remains unaffected by this type of epidemic. About 25% of the United States was affected within 1 year and resulted in a drop of 12 years in United States life expectancy.[9]

SPANISH FLU AND EBOLA VIRUS DISEASE PANDEMICS

There were similarities between the flu epidemic of 1918 and current Ebola virus disease epidemic. The 1918 pandemic mostly killed healthy, young individuals compared to most influenza outbreaks that killed juvenile, elderly, or already weak patients. In this regard, the Spanish flu epidemic shares the same age range of most affected population subset with Ebola virus disease epidemic. Older adults may have had partial protection in 1918 from their earlier exposure in 1889–1890 flu epidemic, but it is unclear whether older adults during the Ebola virus disease outbreak had partial exposure to Ebola virus infection. Both epidemics share the unique vulnerability of pregnant women to infection. The fatality of Spanish flu virus was attributed to cytokine storm, a consequence of excessive stimulation and activity of the immune system.[10] An overactive immune system has been implicated in tissue injury seen in Ebola virus-infected patients as well. Younger patients have a more robust immunological response, which might explain the higher vulnerability and fatality seen in this age group during both the flu epidemic and Ebola virus disease epidemic. Another oddity was that the Spanish flu outbreak was widespread in the summer and autumn, which is a similar pattern seen with Ebola virus infection epidemic. Open quarters and outdoor dwelling permitted by warm weather did not reduce the incidence of either disease.

There were notable differences between the flu epidemic of 1918 and current Ebola virus disease epidemic. The Spanish flu epidemic had the necessary prerequisites for an epidemic. Massive troop movements and close quarters during World War I accelerated the epidemic, probably increasing transmission and the mutation. The virus' lethality might have been increased by war. These soldiers' susceptibility might have been increased by their immune systems weakening through malnourishment, stresses of combat, and chemical attacks.[11] The patterns of influenza pandemics have many atypical features. Pandemic emergence can follow one of these two patterns: de novo emergence of completely unique avian-descended virus or modification of a circulating human-adapted virus by importation of a novel neuraminidase (NA) (e.g., the 1957 H2N2 pandemic) via genetic assortment, of a novel HA hemagglutinin (HA), either with concomitant importation (e.g., the 1968 H3N2 pandemic).[12]

France was at the center of 1918 flu pandemic with a major troop staging and a hospital camp in Étaples identified by a British team's investigative work. This team was headed by a virologist John Oxford from St Bartholomew's Hospital and the Royal London Hospital. The virus was harbored in the birds there that mutated and passed on to pigs, kept near the front line.[13] During World

War I, 96,000 Chinese laborers were mobilized to work behind the British and French lines on the Western Front. This high population density could be the cause of the pandemic. In a recent report, Humphries found archival evidence that a respiratory illness struck Northern China in November 1971, identified by Chinese officials as identical to Spanish flu a year later.[14,15] But the Ebola virus disease epidemic did not have any identifiable predisposing causes.

FINDING THE CULPRIT BEHIND THE PANDEMIC

Unlike the discovery and characterization of the Ebola virus in the 1970s aided by modern marvels such as electron microscopes, the influenza virus evaded researchers for decades. During the 1892 influenza pandemic, German physician Richard Pfeiffer isolated bacteria from lungs and sputum of influenza patients and labeled bacteria as the cause of influenza. Bacteria came to be known as Pfeiffer influenza bacillae and later *Bacillus influenzae* (*Haemophilus influenzae*). However, the controversy continued because strains of streptococcal, pneumococcal, and other bacteria could be found in sputum of patients suffering from the flu, and *B. influenzae* could not be found in samples from many patients with influenza. *Bacillus influenzae* could be found in healthy individuals and in those patients suffering from measles, scarlet fever, diphtheria, and varicella (chicken pox). In one article, the authors wrote "There seems to be no justification for the belief that the epidemic was due to the influenza bacillus, which is probably a secondary invader and bears about the same relation to the influenza cases as to respiratory infections of a different sort."[16]

In 1918, veterinarian J.S. Koen noted similarities of influenza symptoms in pigs while he was treating the symptoms of Spanish flu epidemic in humans. In 1928, researchers N. McBryde and Robert Shope carried out experiments to identify the transmission mechanism of influenza in pigs. They took mucus from infected pigs using bacterial filters to remove any bacteria and then inserted virus-loaded mucus into healthy pigs' nostrils to infect them. McBryde failed, but Robert Shope succeeded in transmitting the infection. His experiments finally proved that the influenza infection was indeed due to a virus.[5]

Finally, in 1933, researchers Wilson Smith, Christopher Andrews, and Patrick Laidrow identified the influenza virus in humans.[5] In the decades to come, the complete genome of the influenza virus would be characterized from a virus derived from the lung tissue of 1918 influenza epidemic victims. The tests showed that the pandemic virus contained genes derived from avian-like influenza virus strain.[4] Almost 90 years later, in December, a University of Wisconsin researcher, Yoshihiro Kawaoka reported that three genes (termed PA, PB1, and PB2) had been identified within the genome of the influenza virus from 1918 (isolated from lung and brain tissue samples of British politician and diplomat, Sir Mark Sykes).[17] These genes enabled the virus to produce three different types of proteins within infected cells, weakening the protective lining of the victim's bronchial tubes and lungs and clearing the way for bacterial pneumonia.

An enzyme, neuraminidase, enhances bacterial growth and bacterial dissemination.[18] A second proapoptotic protein enhances the inflammatory response by unclear mechanisms.[18] And a third protein increased the production of interleukin-10 (IL-10), reducing the function of white blood cells in the lungs.[18]

DEVELOPMENT OF VACCINES AND THE QUAGMIRE OF PSEUDOVACCINES

We have the same challenges in developing a vaccine against Ebola virus that we had against influenza virus. While initial challenges faced by researchers were predominantly due to lack of identification of the appropriate causative organism, this issue will not trouble modern-day researchers. William H. Park, MD, from New York City Health Department, was convinced that *B. influenzae* was the cause of the outbreak and he set about devising a vaccine and antiserum against it. This antiserum was completed on October 17, 1918. In Philadelphia, Paul Lewis worked on refining the pneumococcal vaccines. As a result, Philadelphia municipal laboratory released thousands of vaccines (a mix of streptococcal, pneumococcal, and *B. influenzae*) on October 19, 1918. Physicians at the Naval Hospital on League Island, Pennsylvania (the Philadelphia Naval Shipyard) made the vaccine from *Micrococcus catarrhalis* (now *Moraxella catarrhalis*) and *B. influenzae* and strains of *Pneumococcus, Streptococcus, Staphylococcus*. Each vaccine, administered in four dose regimens, contained 100,000,000 and 200,000,000 bacteria per cubic centimeter. These scientists noted that nonvaccinated individuals and health-care workers became sick even though strict preventive measures like use of masks and gloves were taken. From the group of ill patients vaccinated therapeutically, no one developed pneumonia-like symptoms like those seen in the unvaccinated group. Investigators noted,

> The course of the disease [in those treated therapeutically]…was definitely shortened, and prostration seemed less severe. The patients apparently not benefitted were those admitted from four to seven days after the onset of their illness. These were out of all proportion to the number of pneumonias that developed and the severity of the infection of the control cases. The effects were always more striking the earlier the vaccine was administered.[16]

Finally, they concluded, "The number of patients treated with vaccines and the number immunized with it is entirely too small to allow of any certain deductions; but so far as no untoward results accompany their use, it would seem unquestionably safe and even advisable to recommend their employment."[16]

In San Francisco, another group of researchers studied vaccines use. They mentioned that Spanish influenza did not reach San Francisco until October 1, 1918, therefore, staff at the training station had enough time to prepare a prophylactic vaccine even there was a great debate as to the pandemic's cause.[16] The vaccine was made of *Streptococcus hemolyticus* (*Streptococcus pyogenes*), 100 million, *B. influenzae*, 5 billion bacteria; Pneumococcus types I and II, 3 billion each and Pneumococcus type III, 1 billion bacteria. It was first tested in guinea pigs and then five laboratory volunteers. After vaccination, the serum from the Guinea pigs and volunteers agglutinated when mixed with *B. influenzae* in vitro (meaning that their blood had antibodies that reacted with the bacteria). The vaccine was administered to 11,179 military personnel and civilians, including some at Mare Island (Vallejo, CA), San Pedro, and San Francisco associated with Naval Training station. In vaccinated personnel, the rate of influenza was lower compared to rates seen in unvaccinated persons. The rate of influenza in the control group was 1.5–33.8% and in the vaccinated group was 1.4% and 3.5%.

Another vaccine was used in Washington, USA, at the Puget Sound Navy Yard,[17] where 4212 people were vaccinated with streptococcal bacteria-derived vaccine. The rate of influenza in vaccinated persons ranged from 2% to 57% and 1.8% to 19.6% in those who received the vaccine. Investigators stated, "We believe that the use of killed cultures as described prevented the development of the disease in many of our personnel and modified its course favorable in others."[16] They noted that *B. influenzae* did not play any role in the outbreak.

The use of mixed bacterial vaccine was reported by E. C. Rosenow (Mayo Clinic) in Rochester, Minnesota. In the initial study, three doses of vaccine were given to 21,000 people. He concluded that,

> The total incidence of recognizable influenza, pneumonia, and encephalitis in the inoculated is approximately one-third as great as in the control uninoculated. The total death rate from influenza or pneumonia is only one-fourth as great in the inoculated as in the uninoculated.[16]

He decided to test his vaccine in nearly 100,000 people.[16] The results of such studies, however, were viewed with caution and skepticism. The medicine editors of the *Journal of the American Association* cautioned "the data presented are simply too inadequate to permit a competent judgment" about vaccine effectiveness in an editorial entitled "Prophylactic Inoculation Against Influenza,"[16] in particular, they addressed Rosenow's paper as follows:

> To specify only one case: The experience at a Rochester hospital—where fourteen nurses (out of how many?) developed influenza within two days (how many earlier?) prior to the first inoculation (at what period in the epidemic?), and only one case (out of how many possibilities?) developed subsequently during a period of six weeks—might be duplicated, so far as the facts given are concerned, in the experience of other observers using no vaccines whatever. In other words, unless all the cards are on the table, unless we know so far as possible all the factors that may conceivably influence the results, we cannot have a satisfactory basis for determining whether or not the results of prophylactic inoculation against influenza justify the interpretation they have received in some quarters.[16]

In the 1930s, researchers finally concluded that influenza was, in fact, caused by a virus and not a bacterium. At the University of Michigan, Thomas Francis Jr. and his team of researchers along with U.S. army made a vaccine by using killed cells through stimulating body's specific immunity against influenza virus, two decades later in 1944. The virus loses its virulence when cultured in fertilized hen's eggs showed by Thomas Francis Jr. Earlier work of Frank Macfarlane Burnet became the basis for Thomas Francis Jr.'s effort.[5]

The story of vaccine development may have lessons for investigators involved in vaccine development for the prevention of Ebola virus disease. A 2010 article described the performance of 13 bacterial vaccine studies conducted from 1918 to 1919. The authors concluded that, despite the limited number of bacterial strains in the vaccines, the pneumonia attack rate could have been reduced by some vaccines after viral infection through cross-protection from multiple related strains.[11] Vaccinologist Stanley A. Plotkin, MD, was more skeptical and posits, "the bacterial vaccines developed for Spanish influenza were probably ineffective because at the time it was not known that pneumococcal bacteria come in many, many serotypes and that of the bacterial group they called *B. influenzae*, only one type is a major pathogen."[16] In another sense, vaccine developers had limited ability in identifying, isolating, and producing all potential disease-producing strains of bacteria. Today's pneumococcal vaccine is protective against 13 serotypes, and the adult vaccine protects against 23 serotypes of that bacteria.[16]

There are similar concerns with the Ebola virus that, due to fulminant spread and viral replication, several serotypes of the virus may exist. When DNA is copied in a cell, enzymes called "polymerase" do the building, adding nucleotide after nucleotide until the DNA is copied. Normally, cellular machinery proofreads the DNA, getting rid most of the bad copies, and keeping the mutation rate very low. But, when a virus like Ebola or influenza hijacks the cell to make copies of itself, it uses the host RNA polymerases. RNA polymerases make copies of genes in the DNA in the form of RNA, which is then read by the cell protein factories (ribosomes) to assemble proteins. Unfortunately, RNA polymerases do not proofread well, so lots and lots of mutations slip through. That is the reason flu shots are administered yearly, because the influenza virus (RNA virus) mutates rapidly. The rapid spread of the Ebola virus gives it more opportunity to mutate rapidly. As a result, it becomes hard to treat it. That is why cures and vaccines are difficult to develop when the targets keep changing. Targeting one serotype may not provide the desired protection by vaccination.[19]

ROLE OF GENERAL PREVENTIVE STRATEGIES

There is great interest among health-care professional to prevent and treat the current Ebola virus infection epidemic and to understand general preventive strategies that were effective in 1918–1919 and how they can be applied in current settings. However, such a search may be not a rewarding one.

In Philadelphia after the second wave in late 1918, new flu case appearances dropped abruptly to almost nothing.[10] Most people would like to believe that medical professionals improved strategies in preventing and treating the bacterial pneumonia that developed along with the viral infection, and therefore, the fatality was substantially reduced. In the United States, spitting was banned in public places, and covering mouth was made mandatory while sneezing. Cities and counties began to recommend that citizens should wear masks, but influenza cases continued even in the communities wearing the masks. In 1918, during an epidemic to slow the transmission of the flu epidemic, the New York commissioner ordered that businesses should be opened and closed on staggered shifts to avoid subways overcrowding. People seen spitting on the street were approached by Boy Scouts in New York City who gave them cards reading "You are in violation of the Sanitary Code."[20]

To prevent disease from spreading, quarantine was imposed in many communities in the United States. Theaters, schools, saloons, pool halls, and churches were closed. Some physicians suggested that drinking alcohol could prevent infection, which caused a high demand on supplies of alcohol. Public health officials censored newspapers and simple directives to stop the rising panic about influenza disease, but posters and cartoons were allowed to warn people about influenza. The posters, however, were exclusively printed in English despite knowing the fact that much of the nation's large population did not speak or read English. Even the native speakers found posters and directions confusing. Many folk healers prescribed to wear a special amulet type or a small bag of camphor as preventive measures against influenza. None of these practices proved to be beneficial in preventing a pandemic.[21]

ENCEPHALITIS LETHARGICA (VON ECONOMO'S ENCEPHALITIS) EPIDEMIC (1915–1926)

Encephalitis lethargica literally means "inflammation of the brain that makes you tired."[22] Encephalitis lethargica has also been known as von Economo's encephalitis, lethargic encephalitis, sleeping sickness, epidemic encephalitis, sleepy sickness, von Economo's disease, Schlummerkrankheit, Schlafkrankheit, or simply Economo's disease. Kinnier Wilson and Bernard Sachs named the disease on the basis of the brain region involved such as mesencephalitis and basilar encephalitis, respectively. This illness spread through Europe in a manner consistent with an epidemic, beginning in the winter of 1916–1917. Sporadic appearances of cases continued till 1930.[22]

Doctors Russell Dale and Andrew Church reported that the disease was related to a particular strain of bacteria called diplococcus, a known cause of sore throat. The actual neurological manifestations were thought to be due to exaggerated immune responses to the bacteria particularly targeting neural tissue (cross reactivity). Other researchers, however, believe that the cause of the disease is a viral infection or post-viral disease.[22] There was no evidence that the disease spread by direct contact. The cases were isolated and seemed to appear simultaneously. Most of the physicians did not consider the disease as being contagious. Children and adults were affected and died at the same ratio of 1:2. Only 2% of cases were over the age of 60 years in one report.[23]

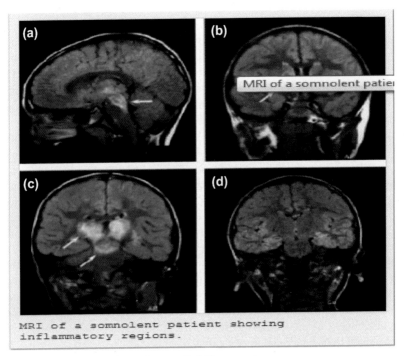

MRI of a somnolent patient showing
inflammatory regions.

Ref. 22.

Young people are most vulnerable to the disease, though it infects people of all ages. The disease typically manifests as sore throat, headache accompanied by fever, double vision, and severe weakness. Within hours, it will progress to episodes of tremors, intense myalgias, involuntary movements, and fatigue. Behavioral changes included psychosis and hallucinations, followed by steadily increasing lethargy and drowsiness. Many patients will eventually become completely unresponsive and comatose. Survivors tended to remain in a vegetative state or coma. Many of those who were affected with the disease, even after some recovery time, continue to experience vision problems, personality changes, difficulty swallowing, and intermittent psychosis.[22] Symptoms and signs of Parkinsonism or persistent catatonia with varying severity of cognitive deficits may be seen in postinfectious period.[22]

The disease was first reported at the beginning of 1917 in Austria by von Economo. One case was noted in each of the months of January, February, and March of that year. In spring of 1918, the disease reached France and England and the following winter, cases were reported in Italy. In February 1919, the disease first appeared in Portugal, in November in India, and in early part of 1920 in Spain. In the United States around the end of 1918, disease first appeared on the Atlantic Coast and in October 1919, on the Pacific Coast. The first case of that year occurred in New York. Subsequently, there were 3 cases in October,

7 cases in November, 10 cases in December 1919; and during 1920, there were 19 cases in January, 35 cases in February, 61 cases in March, 12 cases in April, 5 cases in May, and no reported cases in the month of June. In all of the United States, Public Health Service (PHS) received notification of 222 cases. After the investigation, 39 cases were eliminated as unrelated to the disease.[23]

Cases were reported in South America and Uruguay. The disease was also reported in Australia in March 1919.[23] A short-lived outbreak occurred in three communes of Western Flanders, Belgium, during summer of 1919. In Denmark, from July 1919 to April 1920, 58 cases were reported along with 13 deaths. In England and Wales during 1919, 535 cases were reported, and another 202 cases were reported up to April 10, 1920. Most of the cases occurred during the winter in England, suggesting a seasonal pattern of the disease. A majority of the cases occurred in individuals under the age of 35 years. It was rare for cluster of cases to occur in the same household.[23] In Peru, three cases were reported in 1919. In Poland, cases of encephalitis lethargica were reported in February 1920 when disease notification was made compulsory by the Ministry of Health. Two suspected cases and one confirmed case were reported from Tunis.[23] Since January 1, 1919, encephalitis lethargica has become a notifiable disease in England and Wales.[23]

Netter reported the occurrence of the disease in France at the end of 1918. Since that time, the number of cases started to increase, especially during the winter of 1919–1920 according to French delegate, Dr Pean. In about 2 months, 321 cases were reported out of which 100 died with an estimated case fatality rate of about 30%. During 1918 in Italy, sporadic cases were registered and a considerable number of cases occurred from 1919 to 1920. The greatest number of the cases were reported from the Central and North Italy, while south of the country remained entirely disease-free, suggesting that the maritime climate conferred a certain immunity.[23]

In 1928, the encephalitis lethargica epidemic disappeared with the complete absence of any new cases. Previously affected patients required institutionalization and constant care. In 1969, over 40 years after the occurrence of infection, some catatonic patients showed dramatic improvement in their mobility when treated with levodopa. But recovery was short-lived and most patients deteriorated into catatonic states with repeated or increasing dosages being ineffective.[22]

EBOLA-LIKE HEMORRHAGIC FEVERS IN MEXICO

Another epidemic in the sixteenth-century Mexico bore similar characteristics to the hemorrhagic variant of the Ebola virus disease. In 1545, after 24 years of Spanish rule over the Aztec empire, a disease appeared in the highlands of Mexico that had never been seen before. The disease was characterized by an acute fever, dizziness, severe headaches, bleeding from nose, mouth, and ears along with yellow color of skin (jaundice), and chest pain. The disease persisted for 3–4 days and led to death in a high proportion of the affected persons. The disease only afflicted the native population, sparing the Spanish population for reasons that no one understood until the present day. The Spanish had not experienced any disease like the one that was rampant in Americas, but nonetheless had immunity to the disease. The epidemic of 1545 in Mexico caused about 800,000 deaths in Mexico Valley alone and lasted for 4 years. At that time, the population of Mexico was about 6.6 million. The epidemic was so severe that about 80% of the native population died by the time the disease outbreak had ended.[24]

The disease was named cocoliztli, a word for pestilence in Nahuatl, a Uto-Aztecan language widely spoken in Central and Western Mexico. The epidemic of hemorrhagic fevers was the first in a series of epidemics that devastated Mexico from 1545 to 1815, a period that coincides with Spanish colonial period in Mexico. Such temporal coincidences implicate the Spanish invaders as the harbinger of the disease, although, in the absence of previous exposure, no such association can be confirmed. Another group that could have carried the disease was slaves transported from Africa into the Americas. The inevitable question is whether the disease was a precursor or variant of Ebola hemorrhagic fever seen in Africans in current times. Eleven more outbreaks of cocoliztli were reported, but information about them is scarce. They occurred in 1555, 1559, 1566, 1587–1588, 1592–1593, 1601–1602, 1604–1607, 1613, 1624–1631, 1633–1634, and 1641–1642.[24]

THE EPIDEMIC OF COCOLIZTLI IN 1576

One of the largest epidemics of cocoliztli occurred in 1576 from a population of 4.4 million, causing 2 million deaths in Mexico and a mortality rate to 45% of the entire population. At this time, there were intense changes taking place in Mexico. Diseases like smallpox, measles, mumps, and typhus were taking their toll in the native population. Slaves were being brought from Africa; hard work and high taxes were being imposed on native population; deforestation was intense for the construction of Spanish cities; the Spanish introduced new industries like silver mining, domestic animals, and crops; and constant wars were waged with Indians of the North and conversion to Catholicism was an ongoing process. Under these circumstances, cocoliztli reappeared 31 years after the first outbreak.[24]

The presence of hemorrhagic fever was first reported in June 1576 and quickly became a source of death in just 3 months all over the country. Tepeaca, a city 150-km southeast Mexico City with a preepidemic population of 60,000 was converted into an 8000 postepidemic population and lost 86% of its citizens. Cholula, 95-km east of Mexico City, saw a decline from 15,000 inhabitants to 9000 with a 40% death rate. Nochistlan, a town, 450-km northwest of Mexico City suffered from 66.7% loss in its inhabitants secondary to the disease. The overall mortality caused by this epidemic was a loss of about 2 million people from its original 4.4 million. Young indigenous adults suffered the most during this epidemic. The data from the census of Cholula epidemic indicate that 75% of deaths were among individuals 25 years or older. The infant mortality remained unchanged from previous years. In regards to population segment afflicted, the disease shared features with flu and Ebola virus disease epidemics that came five centuries later. During this epidemic, weather remained cloudy, cold, and foul; the rainy season started 2 months earlier than it was expected and was intense.[24]

Fray Juan de Torquemada, a Franciscan historian, described epidemic of 1576 in Mexico City magnitude as follows.

In the year 1576, a great mortality and pestilence that lasted for more than a year overcame the Indians. It was so big that it ruined and destroyed almost the entire land. The place we know as New Spain was left almost empty. It was a thing of great bewilderment to see the people die. Many were dead and others almost dead, and nobody had the health or strength to help the diseased or bury the dead. In the cities and large towns, big ditches were dug, and from morning to sunset the priests did nothing else but carry the dead bodies and throw them into the ditches without any of the solemnity usually reserved for the dead, because the time did not allow otherwise. At night, they covered the ditches with dirt… It lasted for one and a half years, and with great excess in the number of deaths.[24]

Martin Enriquez wanted to know the number of missing people in New Spain. After searching in towns and neighborhoods, it was found that the number of deaths was more than 2 million.[24]

Dr Francisco Hernandez, the protomedico (physician-in-chief) of New Spain and former physician of King Phillip II of Spain wrote:

The fevers were contagious, burning, and continuous, all of them pestilential, in most part lethal. The tongue was dry and black. Enormous thirst. Urine of the colors sea-green, vegetal-green, and black, sometimes passing from the greenish color to the pale. Pulse was frequent, fast, small, and weak—sometimes even null. The eyes and the whole body were yellow. This stage was followed by delirium and seizures. Then, hard and painful nodules appeared behind one or both ears along with heartache, chest pain, abdominal pain, tremor, great anxiety, and dysentery. The blood that flowed when cutting a vein had a green color or was very pale, dry, and without serosity. In some cases gangrene and sphacelus invaded their lips, pudendal regions, and other regions of the body with putrefact members. Blood flowed from the ears and in many cases blood truly gushed from the nose. Of those with recurring disease, almost none was saved. Many were saved if the flux of blood through the nose was stopped in time; the rest died. Those attacked by dysentery were usually saved if they complied with the medication. The abscesses behind the ears were not lethal. If somehow their size was reduced either by spontaneous maturation or given exit by perforation with cauteries, the liquid part of the blood flowed or the pus was eliminated; and with it, the cause of the disease was also eliminated, as was the case of those with abundant and pale urine. At autopsy, the liver was greatly enlarged. The heart was black, first draining a yellowish liquid and then black blood. The spleen and lungs were black and semi-putrefacted. The bile was observed in its container. The abdomen dry. The rest of the body, anywhere it was cut, was extremely pale. This epidemic attacked mainly young people and seldom the elder ones. Even if old people were affected they were able to overcome the disease and save their lives. The epidemic started in June 1576 and were not over in December, when I am writing these lines. Of all New Spain, the disease invaded cold lands (highlands) in the perimeter of 400 miles, and had a lesser effect in lowlands. The disease attacked primarily regions populated by Indians here and there, then regions of mixed population of Indians and Spaniards, later the Ethiopians, and now, finally the Spaniards. The weather was dry and quiet, and disturbed by earthquakes, the air was impure, filled with clouds but without resolving into rain... Very few with abdominal distention were saved. In the beginning, the blood was expelled by some without severe disease, then by very few. Vital energy was consumed quickly.[24]

Dr Hinojoso made some additional observations, mentioning that the disease lasted 3–4 days from onset to death and that on the second or third day the patient became insane and irritated, eyes were red, thirst was insatiable, and nodules behind the ears and neck were so big that they covered entire neck and half of face. Patient fevers were very high, and autopsies showed extremely large and hard livers and also identified the splenomegaly.[24]

The striking aspect of this epidemic was its selective vulnerability of the native population. The selective vulnerability of the indigenous people was not limited to cocoliztli, but also seen with smallpox, chicken pox, measles, and mumps infections. With those diseases, however, an explanation was more forthcoming. The Spanish population had acquired a protective immunity to these diseases due to exposure in Europe at a young age. The native population developed some degree of immunity to those diseases about 20 years later.[24]

In 1576, the time of the second epidemic of cocoliztli, many of the Spanish settlers who helped the sick natives were then between 0 and 54 years, but none of them died of disease. Why did the Spaniards who grew up in the new colonies continue to manifest immunity seen in those who had come from Europe? Immunity is not inherited and therefore it is unlikely that, 55 years later, the immune status of Spanish immigrants played any role in epidemic selection. A possible explanation for this selection could lie in the socioeconomic status of that indigenous and colonizing Spanish population as the majority of the native population lived in poverty and were undernourished.[24]

Cristobal Godinez, a government official, reporting on the epidemic wrote:

The reason so many Indians die of the pestilence is a God secret. I do not find any better answer than that in the past the Indians were not as badly mistreated and oppressed as they are today with heavy workloads. They are skinny and delicate, the disease finds them overworked and without resistance, so they are finished.[24]

Under those conditions, the pattern of cocoliztli epidemic is well explained by the presence of an infectious agent favored by poor living conditions or nutritional deficiencies.[24]

HOW COCOLIZTLI DIFFERED FROM OTHER MAJOR DISEASES

What was cocoliztli? It was a disease that does not exist today or a variant of the present diseases. Cocoliztli differed from epidemic typhus (tabardete) in that rashes were absent. The lack of respiratory symptoms in cocoliztli resulted in the dismissal of influenza, pertussis, or diphtheria infection. Fever caused by malaria runs in spikes, whereas a cocoliztli fever ran continuously high. Symptoms in intestinal anthrax infections predominantly consisted of gastrointestinal tract symptoms such as diarrhea and vomiting, while in cocoliztli there were no such symptoms. Patients of cocoliztli had severe yellow discoloration of the skin (jaundice), severe bleeding, and injected eyes, which are not classical description of bubonic plague. The absence of rash, high fever, and high prevalence of enlarged liver and liver failure in cocoliztli made the possibility of infection by filoviruses such as Ebola and Marburg virus less likely but certainly a possibility. The true etiological agent for cocoliztli is still unknown. The illness ran uncontrolled and caused devastating damage to Indian population for about a century.[24]

EPIDEMICS IN MODERN TIMES

Severe acute respiratory syndrome (SARS) in Southeast Asia

Introduction

Severe acute respiratory syndrome (SARS) is a highly infectious disease caused by coronavirus. This previously unknown virus was first identified in Asia in early 2003 and named as "SARS-associated coronavirus" or SARS-CoV. Since October 2012, SARS-CoV has been under the care of National Select Agent Registry that monitors the handling and possession of bacteria, viruses, or toxins that have the potential to be a severe threat to public health and safety.[25] The initial cases of SARS appeared in last part of 2002 in Guangdong Province of China. The contagious nature of the disease and delayed public response caused the epidemic to spread around the globe very rapidly.[25]

Disease manifestations varied considerably depending on the age and physical status of the patient. Fever and dry cough are the most common symptoms although not specific to SARS.[1] At the beginning of the third week, young patients especially did improve from infection. About a fifth of all patients would have progressive pulmonary involvement requiring mechanical ventilation and 10% of affected patients would die. Deaths, far more prevalent in those aged over 65 years of age, might be due to SARS virus alone or in combination with secondary infections. Investigators also imply a role of exacerbated immune response in pulmonary injury.[1]

"A walk through the course of the SARS epidemic"

The first known case appeared on November 16, 2002, in Guangdong Province, China. On February 13, 2003, 300 cases including five deaths in Guangdong Province due to "acute respiratory syndrome" were reported by the Chinese Ministry of Health. On March 11, an outbreak of SARS among hospital workers was reported in Hong Kong. On March 15, SARS was confirmed as a "worldwide health threat" by World Health Organization (WHO). Possible cases were identified in Canada, Indonesia, Philippines, Singapore, Thailand, and Vietnam. WHO issued an advisory for travelers going to Southeast Asia about the SARS. By March 19, SARS cases emerged in the USA, UK, Spain, Germany, and Slovenia. On March 27, WHO recommended the screening of travelers arriving from worst-affected areas. On March 29, Carlo Urbani, one of the WHO officials who identified SARS, died from the disease. On April 2, WHO recommended the postponement of all nonessential travel to Hong Kong and Guangdong Province of China. By April, the slow response and release of only selected information by Chinese officials became apparent, leading to a public apology. On April 9 and 17, the first SARS cases were reported in Africa and India, respectively. On April 14, Canadian scientists confirmed that they had sequenced the genome of the SARS virus. On April 23, WHO recommended postponement of nonessential travel to Toronto, Canada.[26]

As the infection spread, more radical measures were put into place. All schools are shut down in Beijing for 2 weeks to halt the spread of the disease. On April 26, 13 East and Southeast Asian countries' health ministers met in Malaysia to call for all international travelers to be screened for SARS. On April 27, all entertainment venues, including theaters, cinemas, and karaoke bars, were ordered by the Beijing authorities to be closed down until the outbreak was determined to be an end. On May 5, 10,000 people were quarantined in eastern city of Nanjing, China, by authorities. On May 11, spitting was banned in public places in southern Chinese city of Guangzhou to control SARS. On May 15, in China anyone who broke the quarantine rules was threatened with execution or imprisonment for life. On May 22, the infection reached its peak in Taiwan, with emergence of 65 new cases in one day.[26]

The first hope of epidemic control appeared on April 28, when Vietnam was deemed to have contained the spread of infection with no new cases reported for 20 consecutive days. On May 31, WHO declared Singapore SARS-free. On June 5, WHO announced that the outbreak had peaked the World including China. On June 13, WHO withdrew its travel warnings for Chinese provinces of Hebei, Inner Mongolia, Shanxi, and Tianjin but maintained the warning for Beijing. On June 17, the travel advisory was removed for travel to Taiwan. On June 23, China and Hong Kong were removed from the WHO list of SARS-infected areas. On July 2, after 20 consecutive days without SARS new cases, Toronto was declared SARS-free by the WHO. On

July 5, Taiwan was the last country to be removed from WHO list of SARS-infected areas.[26]

Sporadic cases of SARS reemerged in Asia. On September 9, Singapore announced a new SARS case, but was never confirmed by WHO. On December 17, a medical researcher at Taipei military hospital contracted the virus, according to Taiwan health officials. On January 5, 2004, a 32-year-old man in southern Guangdong Province had developed SARS as confirmed by Chinese health officials. On January 16, 2004, WHO said that it found the evidence suggesting that civets do carry SARS. Plans were announced to slaughter thousands of civet cats to prevent the spread of the disease. On January 17, 2004, Chinese authorities confirmed two new cases in Guangdong Province. On April 26, 2004, the Chinese health officials said that it was investigating at least four new suspected cases and all new cases are connected to a confirmed patient who worked at a SARS research laboratory. On May 19, 2004, the WHO had declared that China had contained the latest outbreak of SARS.[26]

The political ramifications continued in the months that followed. In July 2004, Li Liming, the director of China's main disease control center resigned over the April outbreak, which happened at one of its laboratories. On July 7, 2004, Yeoh Eng Kiong, the Hong Kong health secretary at the time, resigned after being criticized over his handling of the 2003 SARS crisis. He was accused of paying too little attention to SARS when it first appeared in China and about issuing the misleading statements to Hong Kong public.[26]

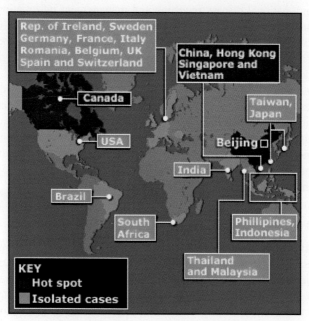

Geographic distribution of SARS occurrence.[27]

FACTORS THAT PROMOTED THE GLOBAL SPREAD OF SARS

There are two main reasons why SARS spread in China. First, is because of the close interaction of Chinese people with wildlife. Second, in many parts of China, people's diets are selenium-deficient and this deficiency may play a role in the emergence of new viral strains. A team led by Melinda Beck of the University of North Carolina at Chapel Hill observed very high mutation rates of flu viruses when they infected selenium-deficient mice. Beck issued a statement saying that "The fact that China has widespread selenium-deficient areas, may play a role in the emergence of new viral strains."[25] SARS spread to Hong Kong when a doctor from Mainland China visited there. Unfortunately, he brought the infection from China, and the infection spread to seven other people who had been staying at the same hotel. Out of these seven people, three were from Singapore, two were from Canada, one was from Vietnam, and one was a local person. Each of these people carried the virus to their respective countries.[28]

STEPS THAT WERE EFFECTIVE IN CONTROL OF SARS EPIDEMIC

Preventive and control measures in China

The government got involved by sending supervisory teams to all 31 provinces to examine the implementation of local control measures. Thus, SARS control measures got incorporated into the legal framework through these legislative efforts. After April 20, 2003, SARS data were reported, analyzed, and managed through national disease reporting and management information system. Close contacts of SARS-infected patients were put under medical observation for 2 weeks to ensure early detection, reporting, isolation, and treatment. In all provinces, special fever clinics and designated SARS hospitals were set up and medical and technical expertise were improved.[29]

A central budget was allocated to treat the farmers and urban residents who had financial difficulties. SARS patients' feces, secretions, dead bodies, and other medical wastes were disinfected diligently. Passenger observation, monitoring, registration, and follow-up systems were set up in civil aviation, railway, long-distance bus, boat, and other public transport systems. The Ministry of Science and Technology coordinated national scientific research into SARS to improve knowledge and understanding of the disease.[29]

Preventive and control measures in Hong Kong

The Department of Health of the Hong Kong Special Administrative Region Government passed legislation making SARS a notifiable infectious disease. SARS patients were isolated in hospitals and family or close contacts were kept under surveillance. Public health workers undertook investigations to identify the source of infection, tracing the contacts, and promoting application of control measures. The government, in May 2003, established three committees: one responsible for the overall cleansing campaigns; one to revitalize city economy including tourism, employment, and trade one to devise ways to promote community involvement in improving the physical, social, and economic environments of the city. The government strengthened collaboration and communication with Mainland China and WHO. Funds were approved to support research on diagnosis, treatment, and SARS vaccine development.[28]

DENGUE FEVER IN ASIA AND SOUTH AMERICA

Dengue fever is a mosquito-borne virus disease that has rapidly spread to several regions in recent times.[30] Dengue fever presents with high-grade fever, headache, mouth nose bleeding, muscle joint pains, vomiting, rash, diarrhea, gastrointestinal bleeding, altered consciousness, seizures, and itching.[31] The disease spread throughout the tropics with local variations, influenced by rainfall, unplanned rapid urbanization, and temperature. In 1950s, during dengue epidemic in the Philippines and Thailand, a severe form, known as dengue hemorrhagic fever, was first recognized, and has become a leading cause of death among children in Asian and Latin American countries.[30] Unlike the Ebola virus infection, which causes a depletion of coagulation proteins by affecting the liver, dengue fever causes hemorrhages by depleting the platelets within the circulation.[32]

About 500,000 dengue fever cases require hospitalization worldwide each year with large proportion of whom are children. Close to 2.5% of those affected with dengue fever will die from the disease.[30] There are four serotypes of dengue fever virus. Recovery from one serotype does not provide complete immunity against other serotypes. A subsequent infection by the other serotype increases the risk of developing dengue hemorrhagic fever.[30]

TRANSMISSION AND SPREAD OF DENGUE FEVER

The virus gets transmitted by female mosquitoes, mainly by the species *Aedes aegypti* and, to a lesser extent, by *Aedes albopictus*. Infected humans are the main multipliers and carriers of the virus, a source of acquisition of the virus for uninfected mosquitoes. After the first appearance of their symptoms, the patients can transmit the infection for 4–5 days (maximum of 12 days) through *Aedes* mosquitoes. The *A. aegypti* mosquito breeds mostly in man-made containers and lives in the urban habitats. It is a daytime breeder and its peak biting periods are early morning and in the evening before dusk. During each feeding period, female *A. aegypti* bites multiple people. *Aedes albopictus* is the secondary dengue vector in Asia.[30]

International trade in products like lucky bamboo led to the spread of *A. albopictus* from Asia to North America and Europe. *Aedes albopictus* is highly adaptable and able to survive in cooler, temperate regions of Europe. It can tolerate temperatures below freezing by hibernation and can shelter in microhabitats.[30]

GLOBAL BURDEN OF DENGUE FEVER

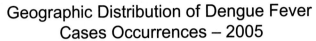

Geographic Distribution of Dengue Fever
Cases Occurrences – 2005

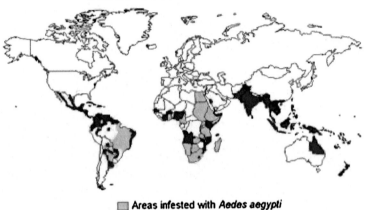

■ Areas infested with *Aedes aegypti*
■ Areas with *Aedes aegypti* and dengue epidemic activity

Ref. 33.

The dengue fever incidence has recently increased dramatically. There appears to be a slowly increasing incidence with intermittent outbreaks consisting of large regional clusters of cases. The actual number of cases is underreported and many are misclassified. One recent estimate shows that 390 million dengue virus infections occur per year, of which 96 million infections have clinical manifestations. In 2013, 37,687 cases of hemorrhagic fever were reported within WHO member states across Americas, Western Pacific, and Southeast Asia, of which 2.35 million cases were related to the dengue virus.[30]

Local dengue fever transmission was reported in France and Croatia in 2010 for the first time, and, in three other European countries, new cases among immigrant populations were detected. In 2012, dengue fever outbreak on the Madeira Islands of Portugal resulted in over 2000 cases. Cases among immigrants were detected in mainland Portugal and 10 European countries.[30] New cases were identified in Florida (United States of America) in 2013 and Yunnan (a province of China). South American countries, importantly Costa Rica, Mexico, and Honduras, also identified new cases. After a lapse of many years, Singapore also reported an increase in new cases, and Laos reported an outbreak. Trend analysis in 2014 indicated an increase in the number of cases in Pacific Island countries notably the People's Republic of China, Fiji, Malaysia, Vanuatu, and the Cook Islands. After a lapse of 70 years, dengue fever was reported in Japan.[30]

THE STORY OF QUARANTINE

Quarantine is a state of enforced isolation used to separate and restrict the movements of persons who may have possibly be exposed to a communicable disease.[34]

The concept of quarantine can be seen as early as the Old Testament. Under the Mosaic Law among the Israelites, as recorded in the Old Testament, infected people were separated to prevent the spread of disease.[35] The quarantine period could be very long such as Mary Mallon (also known as Typhoid Mary, a typhoid fever carrier who spent the last 24 years of her life under quarantine), or it could be very short, such as in the case of suspected anthrax attack (the person is allowed to leave as soon as he or she sheds his or her potentially contaminated clothes and undergoes a decontamination shower).[36]

The Italian words quaranta giorni meaning "40 days" gave origin to quarantine. To protect coastal cities from plague epidemics, this practice began during the fourteenth century. Ships arriving from infected ports to Venice were bound to sit at anchor for 40 days as a quarantine measure.[36]

The Black Death annihilated about 30% of Europe's population along with a significant percentage of Asia's population between 1348 and 1359. Newcomers entering the city had to spend 30 days (a trentine) in an isolated place (nearby islands) waiting to see if the symptoms of Black Death would appear, as stated in the original document from 1377 kept in the Archives of Dubrovnik. Later on, it was prolonged to 40 days, changing the term from "trentine" (30 days) to "quarantine" (40 days).[37]

At the start of Black Death in 1348, three guardians of public health were appointed in Venice, Italy, to check the spread of plague. The next record of preventive measures comes from Reggio in Modena in 1374. Venice founded the first lazaret (quarantine station for maritime travelers) on a small island adjoining the city in 1403. The old leper hospital of Marseille was converted into a plague hospital when Genoa, Italy, followed Venice's example in 1476. Perhaps a complete lazaret of kind, "The great lazaret of Marseilles" was founded on the island of Pomègues in 1526. At all the Mediterranean lazarets, the practice did not differ from the English procedure in the Levantine and North African trade. In 1831 at the western ports, new lazarets were set up during the approach of cholera in 1831, showing the continued use of this system for disease outbreaks.[36]

SYMBOLS OF QUARANTINE

To represent disease, green, plain yellow, and even black flags have been used on ships and ports. The yellow color has longer historical precedent as being a color for marking houses of infection and maritime marking color for disease. The present flag used is the "Lima" (L) flag, mixture of yellow and black flags, also called "Yellow Jack." The disease yellow fever probably derived its name from the flag not from the color of the victim. The plain yellow flag ("Quebec" or Q) probably derived its letter symbol from its initial use in quarantine, but, in present times, it means the opposite, declaring a ship free from quarantinable disease and requesting boarding and routine port inspection. The signal flag "Lima" also called the "Yellow Jack" showed that ship is under quarantine.[36] The simple yellow now indicates that ship is free of a disease.[37,38]

Yellow Jade Flag   Plain Yellow Flag

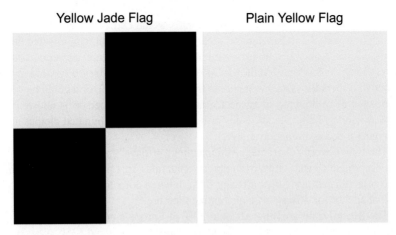

Refs. 36 and 38.

GLOBAL FACES OF QUARANTINE

With the aim of keeping infection out of east and preventing its spread within Europe, many conferences involving the European Powers have been held since 1852. All these conferences were primarily focused on limiting the spread of cholera, but that of 1897. Conferences at Paris (1852), Constantinople (1866), Vienna (1874), and Rome (1885) were fruitless. Each international sanitary convention aimed to make governments follow a uniform set of minimum of preventive actions. The individual countries could have further restrictions.

1. Quarantine rules in Australia

There are many pets and diseases present in Southeast Asia and the Pacific not present in Australia. Due to its proximity to these regions, quarantine is very important in Northern Australia. For this region to protect all Australians, quarantine activation is very important in the region from Cairns to Broome—including the Torres Strait. Due to being isolated geographically for millions of years from major continents, Australia has a distinct ecosystem devoid of many pets and diseases present in other parts of the world. Border inspection of any products that might damage the Australian environment is the responsibility of the Australian Quarantine and Inspection Service. Visitors are bound to fill out the card correctly to declare what food and any other products they bring back with them. Failure to do so will result in quarantine fine of 220 Australian dollars or facing criminal convictions of fining 100,000 Australian dollars and 10 years of imprisonment.[39]

2. Quarantine rules in Canada

Canadian Parliament has passed three quarantine acts: Quarantine Act (humans), Health of Animals Act (animals), and Plant Protection Act (vegetation). In the case of health emergency under the Quarantine Act, the council governor is empowered to block the importation of unnecessary items. If border service officers have reasonable belief that a traveler might be a source of communicable disease or is refusing to answer to necessary questions regarding the public safety, a quarantine officer must be called to isolate that person. In the case of a refusal to be isolated, any peace officer can arrest without a warrant. A quarantine officer who has sufficient belief that traveler might be a source of communicable disease can order treatment after medical examination and detain any traveler who refuses to comply with his or her orders under the law.[40]

3. Quarantine rules in Hong Kong

Health officers may hold the articles he or she believes to be infectious or contain infectious agents under the Prevention and Control of Disease Ordinance (HK Laws. Chapter 599). Failure to submit themselves to a health officer when requested is against the law and will result in arrest and prosecution. The health officer is legally allowed to detain, isolate, and quarantine anyone or anything he or she believes to be infected and restrict any article from leaving the designated quarantine area. Prohibition of the landing or leaving, embarking or disembarking of an aircraft through Civil Aviation Department can also be ordered by him or her.[41]

4. Quarantine rules in the United Kingdom

The quarantine rules in the United Kingdom require that dogs and most other animals must spend 6 months at an HM Customs and Excise pound in quarantine in order to reduce the risk of introduction of rabies from continental Europe. This practice was abolished in 2000 in favor of documentation known as pet passports. Under this scheme, if there is documented proof of animal vaccinations, quarantine can be avoided.[42]

5. Quarantine rules in the United States

The rules imply that, if a disease gets traced back to a particular shipment or product, the United States can put quarantines into effect immediately. In the case of disease outbreak in other countries, all imports will be quarantined. At a number of US ports, small quarantine facilities are operated by the Division of Global Migration and Quarantine (DGMQ) of the Centers for Disease Control and Prevention (CDC). It includes one land crossing (El Paso, Texas) and 19 international airports such as Anchorage, Atlanta, Boston, Chicago, Dallas/Ft. Worth, Detroit, Honolulu, Houston, Los Angeles, Miami, Minneapolis, New York JFK, Newark, Philadelphia, San Diego, San Francisco, San Juan, Seattle, and Washington, D.C. (Dulles). Other ports of entry are also responsible to quarantine potentially infected travelers in their assigned regions.[43,44]

EARLY AMERICAN QUARANTINE

In 1878, continued outbreaks of yellow fever finally pushed Congress to pass quarantine legislation. This paved the way for federal involvement in quarantine activities while conflicting with states' rights.[45]

LATE NINETEENTH CENTURY

Reinterpretation of the law happened in 1892, providing federal government the liberty to impose quarantine requirements following cholera outbreaks from passenger ships arriving from Europe. Control of local quarantine stations was handed over to the US government. Following this, the government built additional federal facilities and increased the number of staff to provide a better coverage. Transferring control of the last quarantine station to the US government in 1921 made the quarantine system fully nationalized.[44]

PUBLIC HEALTH SERVICE ACT

In 1944, the Public Health Service (PHS) Act fully implemented the federal government's quarantine policy for the first time. This gave the U.S. PHS responsibility to prevent the introduction, transmission, and the spread of communicable diseases from foreign countries to the United States.[45]

REORGANIZATION AND EXPANSION

In 1967, quarantine was transferred to an agency known as CDC. The consists of 55 quarantine stations, located at every international airport, port, and a major border station with 500 staff members.[45]

FROM THE INSPECTION TO INTERVENTION

In 1970, the CDC trimmed the quarantine program by changing its focus from routine inspection to enhanced surveillance system based on monitoring the onset of epidemics abroad and modernizing the inspection process for meeting the changing needs of international traffic.[45] In 2003, after the SARS epidemic, the CDC upgraded the quarantine system with 18 stations and greater than 90 field employees.[45]

SOCIAL IMPLICATIONS OF QUARANTINE

Quarantine and other public health practices have always been perceived as intrusive and accompanied by suspicion, distrust, and riots in every age and under all political regimes despite being effective and valuable ways to control the communicable diseases. Standard-Times senior correspondent Steve Urbon describes such temporary quarantine powers, "Civil rights activists in some cases have objected to people being rounded up, stripped and showered against their will. But Capt. Chmiel said local health authorities have 'certain powers to quarantine people.'"[46] Political, economic, social, and ethical issues are always being raised by these strategic measures. Individual rights have often been trampled in the name of public good. The liberty of outwardly healthy persons from lower classes has frequently been violated by isolating or segregating persons suspected of being infected. Marginalized groups of different ethnicities and races have been stigmatized and discriminated against. Quarantine has a line of continuity from the time of plague to the influenza A pandemic in 2009.[46]

During plague and cholera outbreaks, the fear of discrimination and mandatory quarantine pushed the lowest social class and minorities to flee from affected areas. This contributed to a more rapid and wider spread of the disease as this regularly occurred in towns affected by deadly disease outbreak. In this global world, the fear caused by mass media can spread the disease farther and faster, playing a larger role than in the past. Entire populations or segments of populations are at the risk of being stigmatized. In the light of new challenges posed by twenty-first century that can lead to the spread of infectious diseases, quarantine, along with other public health tools, retains importance for public health preparedness. Vigilant attention is required to avoid the intolerance and injustice while implementing these measures. Regular, transparent, and comprehensive communications should be used to gain the public trust. Valuable lessons from the past must lead to successful responses to public health emergencies.[46]

REFERENCES

1. SARS: what have we learned? *Nature* 2003;**424**:121–6.
2. Nadu S. *What are the main causes of Epidemic diseases?* 2015. Available from: http://www. preservearticles.com/201101072771/main-causes-of-epidemic-diseases.html.
3. Chippaux J-P. Outbreaks of Ebola virus disease in Africa: the beginnings of a tragic saga. *J Venom Anim Toxins Incl Trop Dis* 2014;**20**(44).
4. Taubenberger JK. The origin and virulence of the 1918 "Spanish" influenza virus. *Proc Am Philos Soc* 2009;**150**(1):86–112. US National Library of Medicine National Institutes of Health.
5. Panse S. *Who discovered the influenza virus?* 2009. Available from: http://www.brighthub.com/ science/genetics/articles/29659.aspx.
6. King M. *Anoka County History: the influenza epidemic of 1918.* 2014. Available from: http:// abcnewspapers.com/2014/11/13/anoka-county-history-the-influenza-epidemic-of-1918/.
7. *Alfonso XIII of Spain.* 2015. Available from: https://en.wikipedia.org/wiki/Alfonso_XIII_of_ Spain#cite_note-Barry-6.
8. *1918 Flu Pandemic.* 2015. Available from: http://www.history.com/topics/1918-flu-pandemic.
9. *The deadly virus: the influenza epidemic of 1918.* 2015. Available from: https://www.archives. gov/exhibits/influenza-epidemic/.
10. Barry JM. The site of origin of the 1918 influenza pandemic and its public health implications. *J Tanslational Med* 2004;**2**(3).
11. Byerly CR. The U.S. Military and the influenza pandemic of 1918–1919. *Public Health Rep* 2010;**125**(Suppl. 3):82–91. US National Library of Medicine National Institutes of Health.
12. Morens DM, Fauci AS. The 1918 influenza pandemic: insights for the 21st century. *J Infect Dis* 2007;**195**(7):1018–28.
13. Connor S. *Flu epidemic traced to Great War transit camp.* 2000. Available from: http://www. webcitation.org/5ghdTveZz.
14. Humphries MO. Paths of infection: the first world war and the origins of the 1918 influenza pandemic. *War Hist* 2014;**21**:55–81.
15. Vergano D. *1918 flu pandemic that killed 50 million originated in China, historians say.* 2014. Available from: http://news.nationalgeographic.com/news/2014/01/140123-spanish-flu-1918-china-origins-pandemic-science-health/.

16. Youngdahl K. *Spanish influenza pandemic and vaccines.* 2011. Available from: http://www. historyofvaccines.org/content/blog/spanish-influenza-pandemic-and-vaccines.

17. Washington. *Researchers unlock secrets of 1918 flu pandemic.* 2008. Available from: http:// www.reuters.com/article/2008/12/30/us-flu-idUSTRE4BS56420081230.

18. van der Sluijs KF, van der Poll T, Lutter R, Juffermans NP, Schultz MJ. Bench-to-bedside review: bacterial pneumonia with influenza – pathogenesis and clinical implications. *Crit Care* 2010;**14**(2). US National Library of Medicine National Institutes of Health.

19. Mitchell E, Purdom G, Mitchell T. *Is the Ebola epidemic evolution in action?* 2014. Available from: https://answersingenesis.org/biology/microbiology/is-ebola-epidemic-evolution-in-action/.

20. Dominus S. *In 1918 flu outbreak, a cool head prevailed.* 2009. Available from: http://www. nytimes.com/2009/05/01/nyregion/01bigcity.html?_r=0.

21. Services, U.S.D.o.H.a.H. *Fighting Influenza.* 2015. Available from: http://www.flu.gov/ pandemic/history/1918/the_pandemic/fightinginfluenza/.

22. Bellows A. *The sleepy sickness.* 2015. Available from: http://www.damninteresting.com/the-sleepy-sickness/.

23. Health, A.o.S.o.P. World-wide prevalence of Encephalitis lethargica. *Public Health Rep (1896-1970)* 1921;**36**:602–7.

24. Acuna-Soto R, Romero LC, Maguire JH. Large epidemics of hemorrhagic fevers in Mexico 1545-1815. *Am J Trop Med Hyg* 2000;**62**(6):733–9.

25. MedilineNet.com. *What is severe acute respiratory syndrome (SARS)?* 2015. Available from: http://www.medicinenet.com/severe_acute_respiratory_syndrome_sars/page2.htm.

26. News B. *Timeline: SARS virus.* 2004. Available from: http://news.bbc.co.uk/2/hi/asia-pacific/2973415.stm.

27. News B. Spread of SARS worldwide. cited 2015; Available from: http://news.bbc.co.uk/2/ shared/spl/hi/pop_ups/03/health_spread_of_severe_acute_respiratory_syndrome/html/6.stm.

28. Hung LS. The SARS epidemic in Hong Kong: what lessons have we learned? *J R Soc Med* 2003;**96**(8):374–8.

29. Meeting W. *Speech of Mr Gao Qiang, Executive Vice Minister. Ministry of Health, People's Republic of China.* 2003. Available from: http://www.who.int/csr/sars/conference/june_2003/ materials/presentations/qiang/en/.

30. WHO. *Dengue and severe dengue.* 2015. Available from: http://www.who.int/mediacentre/ factsheets/fs117/en/.

31. MayoClinic. *Dengue fever.* 2015. Available from: http://www.mayoclinic.org/diseases-conditions/ dengue-fever/basics/symptoms/con-20032868.

32. Palermo E. *Ebola vs. Hemorrhagic fever: what's the difference?* 2014. Available from: http:// www.livescience.com/48218-ebola-hemorrhagic-fever.html.

33. Images G. *World wide distribution of dengue 2005.* 2005. Available from: https://www. google.com/search?q=World+Wide+Distribution+of+Dengue+2005&client=safari&rls= en&tbm=isch&imgil=-viVDCeSI35qGM%253A%253BRnmqSqskcPI25M%253Bhttp% 25253A%25252F%25252Femedicine.medscape.com%25252Farticle%25252F215840- overview&source=iu&pf=m&fir=-viVDCeSI35qGM%253A%252CRnmqSqskcPI25M% 252C_&usg=__cvp9AXH8MRQq8xDNo14gN6V99PQ%3D&biw=1256&bih=696&ved= 0CCgQyjdqFQoTCPjOgo7OisYCFYg6kgodbY0ADg&ei=bw17VbjbFoj1yATtmoJw#img rc=-viVDCeSI35qGM%253A%3BRnmqSqskcPI25M%3Bhttp%253A%252F%252Fimg. medscape.com%252Fpi%252Femed%252Fckb%252Finfectious_diseases%252F211212- 1641774-215840-2066532tn.jpg%3Bhttp%253A%252F%252Femedicine.medscape. com%252Farticle%252F215840-overview%3B200%3B150.

34. Merriam-Webster. *Quarantine*. 2015. Available from: http://www.merriam-webster.com/dictionary/quarantine.
35. *Leviticus 15*. 2015. Available from: http://www.biblestudytools.com/kjv/leviticus/15.html.
36. Wikipedia. *Quarantine*. 2015. Available from: https://en.wikipedia.org/wiki/Quarantine.
37. Mackowiak PA. The origin of Quarantine. *Clin Infect Dis* 2002;**35**(9):1071–2.
38. Martins-Tuválkin A. *Quaritine Flag*. 2013. Available from: https://flagspot.net/flags/xf~q.html.
39. australia.gov.au. *Customs and quarantine*. 2015. Available from: http://www.australia.gov.au/information-and-services/passports-and-travel/customs-and-quarantine.
40. Canada G.o. *Quarantine Act*. Available from: http://laws-lois.justice.gc.ca/eng/acts/Q-1.1/page-1.html.
41. *Prevention and control of disease ordinance*. 2015. Available from: https://www.ifrc.org/docs/idrl/973EN.pdf.
42. gov.uk. *Pet travel: quarantine*. 2015. Available from: https://www.gov.uk/pet-travel-quarantine.
43. Prevention, C.f.D.C.a. *Quarantine station contact list, map, and fact sheets*. 2015. Available from: http://www.cdc.gov/quarantine/QuarantineStationContactListFull.html.
44. Reports, A.C.R.A.P.A. *Quarantine facilities for arriving air travelers: identification of planning needs and costs*. 2015. Available from: http://onlinepubs.trb.org/onlinepubs/acrp/acrp_rpt_005.pdf.
45. Prevention, C.f.D.C.a. *History of quarantine*. 2015. Available from: http://www.cdc.gov/quarantine/historyquarantine.html.
46. Prevention, C.f.D.C.a. *Lessons from the history of Quarantine, from plague to Influenza A*. 2013. Available from: http://wwwnc.cdc.gov/eid/article/19/2/12-0312_article.

Chapter 5

Most Recent Outbreaks 2014

Chapter Outline

EBOLA VIRUS DISEASE OUTBREAK: GUINEA 2013–2014

Guinea, officially known as the Republic of Guinea, is a West African country that has an estimated population of 10.5 million with a population density of 46.6 persons per square kilometer according to the United Nations 2012 data. The population density of the United States is 35 persons per square kilometer to provide a reference for comparison. The most recent outbreak of Ebola virus infection was suspected to have occurred in a 2-year-old boy who died on December 6, 2013, in a small village in southeastern Guinea called Guéckédou. Further spread of the disease occurred when one or more of the mourners at the funeral got infected and the disease continued on their families and friends.[1]

West African countries have seen their share of epidemics throughout the early and mid-1900s. One of the most iconic of which is the smallpox epidemics that plagued Colonial African countries during those times. The prevention of this epidemic started early in the 1920s, yet it remained a place where smallpox was endemic for a long time. The average cases per year ranged around 15,000–30,000 in Africa from 1928 to 1966. During World War II, Colonial Africa saw another sharp rise in number of smallpox cases. Even by the end of Colonial rule, Guinea still had approximately 1 case per 1000 according to the World Health Organization (WHO) Expert Committee on Smallpox in 1964.[2] Guinea was a region which showed higher prevalence of the disease even though similar climate regions in the same area had lower incidence of the disease.

On March 23, 2014, the Ministry of Health of Guinea had notified the WHO about the emergence of an outbreak of Ebola hemorrhagic fever in forested areas. By March 22, 2014, there were total of 49 cases with 29 of these 49 patients dying from the disease resulting in case fatality rate of 59%. The cases were found in Guéckédou, Macenta, Nzerekore, and Kissidougou districts as well as three suspected cases and two deaths occurred in Conakry. Blood samples were taken from patients and sent to the *Institut Pasteur* in Lyon, France. Ebola virus was detected by polymerase chain reaction (PCR), confirming the virus disease outbreak in Guinea. The Ministry of Health in Guinea with the help of WHO set up multiple measures to help control the spread of the disease. They collaborated with viral hemorrhagic fever laboratory in Donka National Hospital, *Institut Pasteur* in Lyon, the *Institut Pasteur* in Dakar, and Kenema Lassa Fever Laboratory in Sierra Leone.[3] All measures were taken to limit disease spread to neighboring countries like Sierra Leone and Liberia that bordered Guinea districts with confirmed Ebola cases.

The outbreak was declared an epidemic on August 8, 2014. By September 14, there had been an estimated 4507 probable and confirmed cases out of which deaths have occurred in 2296 patients.[4] This outbreak has been larger than all the previously mentioned outbreaks combined in regards to number of persons infected. The disease had already spread to neighboring countries including Liberia, Nigeria, Senegal, and Sierra Leone. The spread may be attributed to the geographical location of the village where first cases were identified. The village is located at the border of Sierra Leone and Liberia with easy access for villagers through borders to both countries.[1]

There are two designated facilities for treatment of Ebola virus disease patients. One is located at Donka National Hospital (Conakry) and second one is at Guéckédou for Ebola virus disease patients located away from Conakry. The Donka Hospital Treatment Facility has an 85-bed treatment center and is primarily managed by *Medecins Sans Frontieres,* which is an international independent medical humanitarian organization that delivers emergency aid to areas struck by epidemics, natural disasters, and conflicts. It was founded in Paris, France, in 1971. Conakry has a population close to 2 million (2014 census) and is thought to contain a quarter of the entire country's population. Guéckédou is an area in Guinea where it is bordered by Sierra Leone (Kenema) and Liberia (Foya) and also houses an 85-bed facility. The population in this area is close to 250,000 (2008 census). The border is close to Kenema and Foya and both these centers have established laboratories. A transit center in Macenta (east of Guéckédou) is a 35-bed center. These patients are eventually transferred to Donka National Hospital or Guéckédou. The transit center has no laboratory. A 30-bed treatment unit in Macenta is currently in progress and should be ready by mid-November and may serve as a third field station, although there are no current laboratories being built for this center for now.

As of December 4, 2014, there have been a total of 2186 cases of Ebola virus disease, according to the Centers for Disease Control and Prevention (CDC). Of these, 1949 cases were confirmed by laboratory tests documenting the presence of virus in the blood. There were 1349 deaths in Guinea attributed to Ebola virus disease.[5]

EBOLA VIRUS DISEASE OUTBREAK: LIBERIA 2013–2014

Liberia is another country situated in the West African region known officially as Republic of Liberia. It borders Sierra Leone to the west and Guinea to the north as well as Ivory Coast to the east. It has an estimated population of 4 million people with a population density of 37.6 persons per square kilometer, according to the 2012 United Nations data. Fear of spread of this Ebola virus disease outbreak led to controlled travel across the 350 miles (560 km) border between Guinea and Liberia. Despite these restrictions, authorities could not control the spread of the disease to Liberia as many people cross each day and strict checks did not catch all the travelers, and many of the roads in and around border are underdeveloped dirt roads. On March 30, 2014, the Ministry of Health in Liberia disclosed information stating that there were seven serum samples from adult patients that had tested positive by PCR to be Ebola Zaire virus. Two of the seven adults had already died.[6]

Most recent news out of Liberia reports that, as of November 2, 2014, Liberia had reported the largest number of cases with a total 6525 infected persons, which is the largest number among all affected countries in West Africa. A total of 2697 deaths have occurred related to Ebola virus disease.[6] The outbreak's earliest recorded case was in March of 2014, when a patient had returned from Guinea and presented with disease in Lofa County, Liberia. Later, catastrophic waves of outbreak broke out in multiple counties in Liberia, peaking in the months of July and September 2014. In June of 2014, the first cases in aggregate were from a densely populated area called Montserrado County with an estimated population of 1.5 million people.

Multiple Ebola virus disease treatment units and Ebola virus laboratory testing areas were set up in and around the counties which had confirmed cases. Such efforts were made possible by a joint collaboration between the Ministry of Health and Social Welfare in Liberia and many other organizations such as WHO, Red Cross, CDC, Samaritan's Purse Liberia, Pentecostal Mission Unlimited, United Nations Population Fund (UNFPA), and United Nations Children's Fund (UNICEF). On September 20, Liberia opened a new 150-bed treatment clinic in Monrovia the capital city for Ebola virus disease patients. By late September, there were about three clinics in Monrovia for the treatment of Ebola virus disease patients. Other clinics set up for Ebola virus infection outbreak control and patient care include a 160-bed facility by Doctors Without Borders in Monrovia; JFK clinic, a 35-bed clinic expanded to 75 beds; and on November 10, the United States opened its first of 17 Ebola virus infection treatment clinics in Tubmanburg.[7–9]

The Ministry of Health and Social Welfare of Liberia also made a task force to reach out to rural areas that were not easily accessible and could have also been case-harboring locations. Many of these areas were not reachable even by road, which required strategies such as preparing teams to visit those locations with proper protective gear and transportation so that quick action could be taken to bring all suspected cases to the nearest Ebola treatment units.[6]

As of December 4, 2014, according to the CDC there have been a total of 7650 cases of Ebola virus disease, out of which 2805 were confirmed by laboratory testing. There had been a total of 3155 deaths in Liberia which was the largest number for any country affected in West Africa.[5]

EBOLA VIRUS DISEASE OUTBREAK: SIERRA LEONE 2013–2014

Sierra Leone is another country situated within West Africa that shares borders with Guinea and Liberia. There is an estimated population of 6 million persons residing in Sierra Leone, according to the 2011 United Nations' estimate with a population density of 82.7 persons per square kilometer. The outbreak in Guinea that occurred in the small village situated on the border lines of Sierra Leone and Liberia that stretches approximately 405 miles (653 km). Up until beginning of May 2014, according to the WHO, there were no reported cases of Ebola virus disease in Sierra Leone. It appeared that Sierra Leone had escaped the initial outbreak.

The outbreak in Sierra Leone is suspected to have started when a traditional tribal healer returned from Guinea after attempting to treat Ebola virus disease patients. He would develop Ebola virus disease and succumb to the disease.[10] The traditional burial methods in West Africa are to wash the body and perform other rituals which involve touching and kissing the body before burial. This particular practice is thought of as the reason for spread of the disease in villages in Sierra Leone. By the end of May, officials had reported 16 cases of Ebola virus disease and 5 of these cases had died. The outbreak in Sierra Leone was reported in a town called Kailahun, capital of Kailahun District in the Eastern Province of Sierra Leone that borders Guinea and is located not far from the site of the initial outbreak that occurred in Guinea.[11]

By July 11, 2014, the disease spread further when it was diagnosed in a patient who traveled from Kenema which is considered the largest city in the country. Further cases were identified in Eastern Province up to the capital of Sierra Leone, Freetown. The health ministry reported in July 2014 that there had been 99 deaths related to Ebola virus disease in Sierra Leone, among which 315 patients were confirmed to have the Ebola virus infection.[12] It was at that time, a doctor considered a hero among health-care workers in Sierra Leone fell sick and was diagnosed with Ebola virus disease while treating patients with the hemorrhagic fever. One of the most controversial events occurred when he was rushed to an Ebola treatment unit run by *Medecins Sans Frontieres,* who had the only known potential treatment at the time for Ebola, ZMapp. However, the medication was not given to the "hero doctor" Sheik Umar Khan and he subsequently died on July 29, 2014.[10]

The situation in Sierra Leone could be described as the worst of the current outbreak. As doctors desperate for basic supplies and medicine battled the deadly disease, health officials were only able to isolate patients waiting to travel to official treatment centers such as the one located in Kenema. The center was started by the United Kingdom, and the United Kingdom government will be assisting Sierra Leone by building four new Ebola virus infection treatment facilities near urban centers.[13] On July 12, 2014, Médecins Sans Frontières stated that they have treated more than 70 patients with symptoms of the deadly virus in an Ebola virus disease treatment center located in eastern Sierra Leone town called Kailahun. To further assist in the large-scale inpouring of patients, they have also expanded the clinic from 32 to 65 beds.[14] According to the WHO Ebola Response Roadmap Situation Report, on September 21, 2014, Sierra Leone needs an additional 532 beds to treat the overwhelming number of Ebola virus disease cases.[15]

As of December 4, 2014, according to the CDC, there have been total of 7420 Ebola virus disease cases, out of which 6039 have been confirmed by laboratory testing. The proportion of patients with confirmed Ebola virus disease is higher than other countries where a large proportion of cases are suspected but without laboratory test confirmation. A total of 1609 patients have died due to Ebola virus disease.[5]

EBOLA VIRUS DISEASE OUTBREAK: SENEGAL, NIGERIA, AND MALI (2013–2014)

Other countries in and around West African region that have been affected with confirmed cases are Nigeria and Mali. Nigeria, according to the CDC as of December 4, 2014, has had 20 cases with the diagnosis of Ebola virus infection; confirmation was by laboratory tests in 19 cases. 8 of the 20 cases died as a consequence of the disease. Mali has seen a total of eight cases with Ebola virus disease and six of them have died.[5]

Senegal, another West African country, has one of the largest populations in the region with a total of 13 million people. Senegal authorities reported to WHO of its first and the only case of Ebola virus disease on August 29, 2014. The blood test, which confirmed the diagnosis of Ebola virus infection, was performed in a laboratory at the *Institut Pasteur* in Dakar, the capital city. The patient was a 21-year-old man who was a native of Guinea arriving in Dakar on August 20, 2014. He sought medical care for typical symptoms suggestive of Ebola virus disease on August 23, 2014. On August 27, the authorities at Conakry in Guinea sent out an alert about a person who was in close contact with a confirmed Ebola virus disease patient and had escaped. This patient was the only reported and laboratory confirmed case in Senegal to date.[16] So far it seems that Senegal has been the luckiest country dodging the Ebola virus disease bullet with a single case and no deaths despite the fact that the patient spent time with his family in Senegal.

INTERNATIONAL SPREAD OF EBOLA VIRUS DISEASE 2014

According to WHO, the CDC, and Doctors Without Borders, the Ebola virus disease has been exported to countries beyond the African continent into countries including the United States, France, Britain, Spain, Germany, and Norway, none of which have ever experienced Ebola virus disease outbreaks. Most infections occurred in health-care workers who had been exposed to the disease in West Africa.[1]

REFERENCES

1. Denise Grady SF. Tracing Ebola's breakout to an African 2-year-old. *NY Times* 2014. [cited 02.12.14]. Available from: http://www.nytimes.com/2014/08/10/world/africa/tracing-ebolas-breakout-to-an-african-2-year-old.html?_r=2.
2. Schneider WH. Smallpox in Africa during colonial rule. *Med Hist* April 2009;**53**(2):193–227. PubMed PMID: 19367346. Pubmed Central PMCID: PMC2668906. Epub 2009/04/16. eng.
3. *Ebola virus disease in Guinea*. World Health Organization; 2014. [cited 25.11.14]. Available from: http://www.afro.who.int/en/clusters-a-programmes/dpc/epidemic-a-pandemic-alert-and-response/outbreak-news/4063-ebola-virus-disease-in-guinea.html.
4. Ebola virus disease in West Africa – the first 9 months of the epidemic and forward projections. *N Engl J Med* October 16, 2014;**371**(16):1481–95. PubMed PMID: 25244186. Pubmed Central PMCID: PMC4235004. Epub 2014/09/23. eng.
5. *2014 Ebola outbreak in West Africa – case counts*. Centers for Disease Control and Prevention; 2014. [cited 12.10.14]. Case Counts. Available from: http://www.cdc.gov/vhf/ebola/outbreaks/2014-west-africa/case-counts.html.
6. Nyenswah T, Fahnbulleh M, Massaquoi M, Nagbe T, Bawo L, Falla JD, et al. Ebola epidemic – Liberia, March–October 2014. *MMWR Morb Mortal Wkly Rep* November 21, 2014;**63**(46): 1082–6. PubMed PMID: 25412068. Epub 2014/11/21. eng.
7. Beds scarce, staff scarcer, in Liberia's overrun Ebola wards. *Jam Obs* 2014. [cited 12.10.14]. Available from: http://www.jamaicaobserver.com/news/Beds-scarce–staff-scarcer–in-Liberias-overrun-Ebola-wards.
8. Karmo H. *Liberia: JFK Ebola unit beyond capacity – patients lying on floor: allAfrica*. 2014. [cited 12.10.14]. Available from: http://allafrica.com/stories/201409261258.html.
9. Paye-layleh JUS. Opens first of 17 Ebola treatment units in Liberia. *Huffington Post* 2014. [cited 12.08.14]. Available from: http://www.huffingtonpost.com/2014/11/10/us-ebola-unit-liberia_n_6134082.html.
10. Umaru Fofana DF. Sierra Leone 'hero' doctor's death exposes slow Ebola response. *Reuters* 2014. [cited 12.06.12]. Available from: http://in.reuters.com/article/2014/08/24/us-health-ebola-khan-insight-idINKBN0GO07M20140824.
11. Schnirring L. *Ebola expands in Guinea, sickens more in Sierra Leone*. Center for Infectious Disease Research and Policy; 2014. [cited 12.02.14]. Available from: http://www.cidrap.umn.edu/news-perspective/2014/05/ebola-expands-guinea-sickens-more-sierra-leone.
12. Gbandia S. Ebola spreads to Sierra Leone capital of freetown as deaths rise. *Bloomberg* 2014. [cited 12.03.14]. Available from: http://www.bloomberg.com/news/2014-07-12/ebola-spreads-to-sierra-leone-capital-of-freetown-as-deaths-rise.html.
13. O'Carroll L. Ebola: Sierra Leone hospitals running out of basic supplies, say doctors. *Guard* 2014. [cited 12.05.14]. Available from: http://www.theguardian.com/world/2014/oct/03/ebola-sierra-leone-epidemic.

14. *The race to control Ebola in Sierra Leone: Medecins Sans Frontieres—doctors without borders.* 2014. [cited 12.05.14]. Available from: http://www.doctorswithoutborders.org/news-stories/field-news/race-control-ebola-sierra-leone.

15. *Ebola response roadmap – situation report.* World Health Organization; 2014. [cited 12.05.14]. Available from: http://www.who.int/csr/disease/ebola/situation-reports/en/.

16. *Ebola virus disease – Senegal.* World Health Organization; 2014. [cited 12.10.14]. Outbreak News: Available from: http://www.afro.who.int/en/clusters-a-programmes/dpc/epidemic-a-pandemic-alert-and-response/outbreak-news/4265-ebola-virus-disease-senegal.html.

Chapter 6

West Saharan Response to Ebola Virus Disease Epidemic

Chapter Outline

A SMALL VILLAGE

Rolling hills, green pastures, and cozy huts dot the landscape. Farmers tend to their crops and shepherds herd their sheep. Lush green forests are not far from this quaint and peaceful town. Triangularly shaped with rounded edges, some would say that this town is actually an oasis of human civilization in the midst of the dense forests that surround it from almost all directions. It is a picturesque landscape that brings to mind the pastures of a small and thriving European village from the early times. This village, though, is not a European tourist destination where people from bustling metropolises from all over the world come to marvel at a country life they fancy in some distant daydreams. Getting to the destination is no easy task. This is the once relatively unknown town of Meliandou in the southeastern part of the West African nation of Guinea (Figure 6.1).

As picturesque as it may seem, this very town would transform, over the course of a few weeks to a few months at the end of 2013 and the beginning of 2014, into a town that resembles the fictitious and horrific abandoned town depicted in the movie *The Hills Have Eyes*. Some of its inhabitants would be shunned by outsiders and fellow inhabitants, not because they are mutants attacking normal people, but because they are at the epicenter of the worst Ebola virus disease epidemic to date, and because these villagers would appear to the rest of the world as violent attackers of health-care workers. Accentuating its infamy is the fact that here the borders of Guinea, Liberia, and Sierra Leone come together[1] making Meliandou the prime location for the spread of the wildfirelike epidemic (Figure 6.2).

FIGURE 6.1 A satellite image from Google Earth of Meliandou in the Guéckédou Prefecture of Guinea.

FIGURE 6.2 Satellite image, acquired from Google Earth, of the village of Meliandou, the surrounding hills, and the Guinea–Liberia border.

A burning question in many people's minds has been, "How did it all begin?" The unfortunate truth is that, to date, no one knows how the devastating epidemic really came to life. Meticulous research published in the New England Journal of Medicine has revealed, what medical science refers to as, "patient zero" or the first person who contracted the virus.[2] Yet there are no clues as to how he acquired the infection.

THE RADIO'S LAST BROADCAST

It was an ordinary day, like any other, on December 2, 2013. Emile Ouamouno, a 2-year-old boy, could probably be found enjoyably listening to a portable radio—bright red in color, which his father would later longingly describe as a favorite past time of the 2-year-old.[3] Not far from Emile, we would likely see his older sister, Philomène, dancing around and playing with a ball—intriguing Emile to join her in her games.[3] For Etienne, the father of Emile and Philomène, this was to become the beginning of the most devastating of his days. That day,

Emile developed a fever, black stools, vomiting, and bleeding.[1] In a matter of 4 days, on December 6, 2013,[a] 2-year-old Emile passed away.

To the distress of Etienne, a week later, on December 13, 2013, the mother of the two children passed away after showing signs of bleeding.[2] Two weeks and two days later, Philomène, the 3-year-old sister, passed away after having developed fever, vomiting, and diarrhea 4 days earlier. In a matter of weeks, Meliandou was devoid of the laughter of Emile and Philomène playing outside with their ball or the bright-red portable radio that Emile had taken a liking to. Closely following the death of Philomène, on January 1, 2014, Emile's grandmother also died after having experienced the same symptoms.[2]

Instead of celebrating the beginning of a new year and talking about resolutions and happily anticipating the year ahead, the people of Meliandou gathered to mourn the loss of another life in Etienne's family. At this point, no one knew what was ravaging the town and decimating so many members of the same family. Perhaps, the world would shrug it off as a minor illness whose effects had been accentuated by the prevalent poverty and lack of health-care accessibility and awareness.

From there, on January 29, 2014, a nurse who had been at the funeral of Emile's grandmother began having fever, diarrhea, and vomiting, and, on February 2, 2014, she also passed away.[2] A midwife at the village, began having a fever on January 25, 2014. She was hospitalized in the Guéckédou Prefecture and died there on February 2, 2014.[2] Before succumbing to the fatal disease, this midwife would pass on the virus to the people of her village, Dandou Pombo— also in the Guéckédou Prefecture, and to the health-care worker treating her, resulting in six deaths.[4] At the funeral of Emile's grandmother, there were two other people who contracted the illness and spread it to Dawa, another village in the Guéckédou Prefecture. Then, once the floodgates of the disease opened, there was no way to contain the ravaging waves of death that followed.

DURING THE DEADLY MARCH WE FIND HEROES

Doctors in Guinea began searching for the cause. Doctors Without Borders (*Médecins Sans Frontières*) were already in the country dealing with a malaria outbreak. They realized that this disease outbreak was out of the ordinary and contacted an expert in Geneva. The expert suspected something sinister such as the Marburg virus or the Ebola virus.[1] Guinea's Ministry of Health did not stand idly by. Samples were sent to *Institut Pasteur* (Pasteur Institute) in France. The result: the causative agent was indeed Ebola virus as confirmed by polymerase

---

[a] It is important to note that while the article published in the *New England Journal of Medicine* gives this timeline, the World Health Organization (WHO)'s Global Alert and Response (GAR) Web site begins the timeline on December 26, 2013 (http://www.who.int/csr/disease/ebola/ebola-6-months/guinea/en/).

chain reaction tests on six of seven samples.[5] To add to the fear, this was the Zaire strain, which is the most lethal strain of Ebola virus.[1]

On March 23, 2014, the Ministry of Health of Guinea notified the World Health Organization (WHO) of an outbreak of Ebola virus disease.[6] The outbreak was considered the largest outbreak of Ebola virus disease by June 18, 2014.[7] On August 8, the WHO declared the epidemic to be a "public health emergency of international concern."[6] The epidemic was declared on August 8, 2014. By September 14, there had been an estimated 4507 probable and confirmed cases and 2296 deaths.[6] This outbreak has been larger than all the previous outbreaks combined, which has also caused the spread of the disease to neighboring countries, namely Liberia, Nigeria, Senegal, Sierra Leone, and Mali. The spread may be attributed to the geographical location of the Guéckédou Prefecture, as it is located close to the border of Sierra Leone and Liberia with easy access through borders to both countries.[8]

The most problematic cases of the disease came from Liberia. This was not because of any difference in the strain but because of the manner in which the patients were discovered as patients of Ebola virus disease. One was the late Thomas Eric Duncan who presented symptoms in Texas—the first case to be diagnosed in the United States. The second problematic Liberian case actually became the first case in Nigeria and thus, what medical science calls, the index case of Nigeria. His name has been written in infamy: Patrick Sawyer. A dual citizen of the United States and Liberia, Sawyer had flown into Nigeria, "already sick" and "he should never have been allowed on the plane," according to an online *BBC* article.[9] An ambulance had brought him from the airport as he had collapsed on the tarmac after "getting sick and vomiting on a flight from Liberia."[10] Sawyer was a "Senior Diplomat from Liberia," according to meticulous medical records written at First Consultant Hospital in Nigeria, and published in an online *allAfrica* article.[10] Sawyer had left Liberia to attend a meeting of the Economic Community of West African States (ECOWAS) in the coastal town of Calabar, Nigeria.[11]

Health-care workers in Nigeria had never experienced a case of Ebola virus disease and, consequently, the disease was not first on the list of possible diagnoses. Instead, Sawyer was thought to have malaria, a common ailment in Africa.[11] It was at this critical stage of the disease process in Nigeria, when a national "everyday hero" was born by many accounts.[12] Dr Ameyo Adadevoh saw Sawyer during her rounds the day following his admission.[12] She noted that while laboratory tests confirmed malaria, the patient was not responding to treatment and instead began developing hemorrhagic (bleeding) symptoms.[13] Dr Adadevoh immediately questioned him about exposure to the Ebola virus disease. According to *to The Guardian*, Sawyer "denied contact with an Ebola patient."[12]

At this point in the story, we can find conflicting reports of whether Sawyer was aware of the fact that it was the Ebola virus disease that had caused the demise of his sister, who he had been taking care of in Liberia before arriving at First Consultant Hospital in Nigeria. According to the *New Republic* online magazine, Sawyer was cognizant of the fact that his sister had the Ebola virus disease.[10]

However, an online *CNN Health* article paints a different picture: "Sawyer had been caring for his Ebola-stricken sister in Liberia ... though at the time he didn't know she had Ebola."[14] Interestingly, the account given *in The Guardian* lets the reader read between the lines, "the Liberian had denied contact with an Ebola patient, even though his sister had died of the virus barely two weeks before his arrival in Nigeria."[12] If the majority opinion could be used to answer this question, we might be left with no choice but to agree with the African online publication, *Vanguard*, "Like Sawyer, Duncan also tried to conceal the fact that he might have been infected with the disease from medical authorities who sought information on this from him."[15]

Such discrepancies, however, should not detract us from highlighting the heroism of Dr Adadevoh and the difficult disease process that Sawyer was suffering. Perhaps his disease had advanced to the stage of confusion and irritability and he could not hold himself back from acting irrationally. According to Jonathan Cohn, "at one point, Sawyer ripped out his intravenous lines as he tried to get away, forcing Adadevoh and her staff to restrain him physically in order to keep him in place."[10] The struggles of Dr Adadevoh were not to end there. She had ordered a test for the Ebola virus disease and it had returned positive. Sawyer was not willing to accept the diagnosis and refused quarantine. Dr Adadevoh remained steadfast and succeeded in keeping him under her care. She went on to contact the local health authorities and prepared for the long road ahead. In fact, some accounts describe how the Liberian government officials got involved and began pressuring Dr Adadevoh to discharge Sawyer.[11] "The Liberian ambassador started calling Dr Adadevo, putting pressure on her and the institution" reported an online *BBC* article.[9] Dr Adadevoh refused.

In the end, her determination and her clinical acumen made all the difference in containing the spread of the Ebola virus disease in Nigeria. Unfortunately, Sawyer did not survive and died on July 25, 2014, 5 days after he had been admitted to First Consultant. The damage had been done. Dr Adadevoh and the other health-care professionals who had helped Sawyer were now in danger as they had attended to Sawyer before they knew that he had contracted the Ebola virus disease. An online *BBC* news article reported that "Dr Adadevoh and eleven of her colleagues caught the virus."[9] All the cases of Ebola virus disease in Nigeria can be traced back to Sawyer and this tremendously helped in containing the spread of the disease. In an online article on the *New Republic* Web site, Jonathan Cohn, after complaining of not being able to rely on the US journalists, sums up the disease-controlling effects of isolating Sawyer,

The decision to isolate Sawyer likely prevented him from spreading the disease to many more people in Lagos, a city of 18 million people with large slums. An outbreak in that environment is the scenario public health officials fear most. Tracing known contacts and isolating them—the time-tested strategy that Nigerian and U.S. officials later used to contain the disease—would have been much more difficult, maybe even impossible, in such a dense and disorganized place.[10]

Of the 20 cases in Nigeria, there were 8 fatalities. On August 19, Dr Adadevoh succumbed to her illness and became one of the eight. She left behind a legacy of hard work, dedication, and determination in the fight against the Ebola virus disease. Other factors that aided Nigeria in achieving an adequate response to the epidemic were the prior establishment of disease control organizations such as the Nigerian Center for Disease control (NCDC) and the Disease Surveillance and Response Program of 1998, and previous experience in managing Lassa fever, another viral infection that leads to bleeding.[16]

In the midst of tracing "patient zero" and the "index case," we lose track of the fact that these patients are people as well. At times, the media can inadvertently demonize them. We can only surmise what was going through Sawyer's mind—after all, he was a diplomat and we would expect him to have some understanding of the serious nature of the disease. Some reports claimed that he was searching for a "cure from one of the so-called miracle pastors" in Nigeria.[9] Given the mysterious nature of the disease, the loss of his sister to the disease, and the knowledge that there is no cure, one can imagine how even a well-educated person could decide on the most seemingly inappropriate course of action. The online *CNN Health* article was one of the few, which tried to humanize him. Sawyer was one stop away from "heading home to Minnesota to celebrate his daughters' birthdays." The 40-year-old was survived by 3 young daughters, aged 1, 4, and 5 years.[14]

NO COUNTRY WILL BE EBOLA VIRUS-FREE

The sixth West African country to confirm a case of Ebola virus disease was Mali.[17] Similar to Nigeria and Liberia, the index case was not a patient indigenous to Mali; like Guinea, the first infected person in Mali was a 2-year-old child, Fanta Condé.[18] Fanta and her family lived in Beyla, Guinea, and an incredible story surrounds her arrival to Mali. While in Guinea, a large number of her family began dying of unknown causes. The first of these was her father, who was a health-care worker at a private clinic and a Red Cross volunteer.[19] Fanta's father had been in contact with a farmer who had visited the clinic with his two daughters, all of whom died of undiagnosed causes. When Fanta's father fell ill, the residents of Beyla shunned him and attributed his sickness to witchcraft.[19] As a result, he left Beyla and returned to his native village, where, on October 3, 2014, he died. Back in Beyla, the paternal grandfather of Fanta lost his wife to an unknown disease. On October 9, 2014, two of his other sons became ill and were referred to an Ebola transit center set up by Doctors Without Borders (*Médecins Sans Frontières*) in Macenta. Blood samples taken from these two tested positive for Ebola virus infections. Unfortunately, both brothers died, and one on the way to the hospital.[19]

On October 17, 2014, the paternal grandfather's blood samples tested positive for the Ebola virus infection after he visited a hospital for treatment of rheumatoid arthritis. On October 20, 2014, he died at the Ebola virus disease treatment center in Guéckédou. Following the earlier deaths, a relative living in Kayes, Mali, decided to visit Beyla, Guinea, to offer her condolences. This was a maternal grandmother of Fanta, the index case of Mali. On October 19, the grandmother decided to travel back to Mali and took the 2-year-old child along with her.[19] Fanta was exhibiting what the WHO site refers to as hemorrhagic symptoms but no vomiting or diarrhea. To make matters worse, the child and grandmother traveled some 1200 km (approximately 746 miles) through Mali on public transport.[19] Once in Kayes, the grandmother and Fanta sought care from two traditional healers. It was the second healer, who referred them to a retired nurse, and it was the nurse who suspected the Ebola virus disease as the cause of the child's 40 °C fever.[19]

Fanta was admitted to the hospital in Kayes, and on October 23, she was diagnosed with Ebola virus disease. Despite being placed in isolation and receiving care, this 2-year-old died on October 24, 2014, and became the index case of Mali. However, the epidemic was better controlled in Mali as Malian health officials, the WHO, US Centers for Disease Control and Prevention (CDC), Doctors Without Borders (*Médecins Sans Frontières*), the International Federation of Red Cross and Red Crescent Societies, and a number of other organizations worked diligently to trace the contacts of Fanta.[19] A total of 108 contacts were identified: 33 were health-care workers. As of November 2014, 25 of the 108 were followed for 21 days and were released from the surveillance system.[19]

Unfortunately, right when Mali was getting ready to declare that it was free of the Ebola virus disease outbreak, a second outbreak occurred and it was seen as "larger and more threatening."[18] The second outbreak was discovered when a nurse at the Pasteur Clinic, not associated with the Institut Pasteur, fell ill. On November 10, 2014, doctors ordered an Ebola virus detection test on the nurse and discovered that she had contracted the Ebola virus disease. Her contact was traced back to a religious leader from Kourémalé, a town straddling the Guinea–Mali border.[18] This man had fallen ill on October 17, 2014, and had traveled to Mali for better treatment. It was at the Pasteur Clinic that he was found to have kidney failure and the nurse came into contact with him. Sadly, the private clinic failed to diagnose Ebola virus disease as the underlying cause of his kidney failure and thus, the Ebola virus was granted access to the nation of Mali once again.[18]

Situations like these led people to consider various steps to prevent the import and spread of disease. Closing the border of Mali was one such consideration. However, long traditions of hospitality and welcoming strangers (called *diatiguiya* and pronounced JAH-tih-GEE), along with the repercussions of the arbitrary French colonial division of nations, made closing of borders an option not worth the time or effort.[18] Other efforts to stem the spread of the disease are ongoing. At the forefront of the effort in Mali, as in the other West African countries, has been Doctors Without Borders (*Médecins Sans Frontières*). In coordination with Mali's national disease center, CNAM, Doctors Without Borders (*Médecins Sans Frontières*) has been training local health-care workers in the management of Ebola virus disease patients. It also oversees implementation of ambulance services to transport patients and organizes safe burials.

As time progressed, many more people in Liberia, Sierra Leone, and Sengal contracted the Ebola virus disease and succumbed to its effects. Just in the week of September 14, 2014, Sierra Leone had 210 cases, Liberia had 400 cases, and Guinea had 90 cases of Ebola virus disease.[6] There was also a greater disturbing rise in cases in Liberia over the course of the epidemic as compared to Guinea and Sierra Leone.[6] As of February 9, 2015, Liberia had the highest number of reported cases (8864) despite the fact that in August 2014 the Liberian Ministry of Health and Social Work (MOHSW) had enlisted the help of a number of international organizations including the CDC and the WHO.[20] There are various possible explanations for the rise of cases in Liberia: the nation's health-care workers were already frustrated and had gone on strike, the level of preparedness for any major disease outbreak was dismal, and initially the government and people were in denial.[21] In addition, there are many areas that are hard to access and thus led to unreported and undiagnosed cases. These villages were prone to more outbreaks as a result of "travelers from affected areas (such as Monrovia) returning to their rural homes" and infecting other people, all of whom would remain undetected until suffering fatal consequences.[22] Geleyansiesu is a prime example of one such village; it is only accessible by "canoe and several hours walking."[22] The progression of the Ebola virus disease was rapid in most of the regions. According to the WHO, as of February 15, 2015, there were "a total of 23,253 confirmed, probable, and suspected cases of Ebola (virus disease) and 9380 Ebola (virus disease)-related deaths from the three West African countries (Guinea, Liberia, and Sierra Leone) where (the) Ebola virus (disease) transmission has been widespread and intense."[23] At the onset of the epidemic, Meliandou alone had become the site of 14 graves. At the epicenter of it all, one can only imagine the pain that Etienne had to endure after the death of his son, wife, daughter, and, then his children's grandmother and the potential stigma that might have been created around him.

BURN THEIR MATTRESSES, BURN THEIR THINGS!

While we, on the outside, were able to read about the reaction of the local people to outsiders (the foreign health-care workers or educators visiting them), there was little, if any information, on how the people of Meliandou, and the other towns in the Guéckédou Prefecture where the Ebola virus disease first spread, were reacting to the virus and to the infected people within their own towns. One online *CNN Health* article reports that "Residents, out of fear, also burned… mattresses and other possessions," of the people who had died of Ebola, and that they rejected "the infected children and the other infected family members."[4] This was in stark contrast to their strong tribal cultures and customs, where extended family was known to take in orphans.[4] In fact, according to Fassou Isidor Lama, a Child Protection Officer of UNICEF, as a result of the virus, one could see "people flee their villages, and abandon their families and their children."[4] Fear, suspicion, death, and misery had become rampant in Meliandou and the other towns of the Guéckédou Prefecture. Even local health-care employees were frightened and the hospital director of a treatment center in Guéckédou had to "persuade his employees to come to work."[8]

FIGURE 6.3 A child stops on a Monrovia street Friday, September 12, to look at a man who is suspected of suffering from Ebola virus disease. *From CNN World.*[37]

FIGURE 6.4 A body, reportedly a victim of Ebola virus disease, lies on a street corner in Monrovia on Saturday, August 16. *From CNN World.*[37]

In Liberia, Ebola virus-infected people could be seen lying in the streets; sometimes bodies of people who had died of Ebola virus disease could be seen abandoned in the streets. This fear was only to grow and transfer, paradoxically, against outside health-care workers and educators working to stop the epidemic (Figures 6.3 and 6.4).

BURN THEM AT THE STAKE!

An unwritten and unspoken axiom is that if someone comes to help you out of a hole, you do not turn around and hurl stones at them. However, just like the peculiar and unpredictable nature of this Ebola virus disease epidemic, this axiom was turned on its head in West African Sahara, where the epidemic was wreaking havoc. Contrary to the image presented in the media, the villagers' reactions were not based on a barbaric fear of health-care workers or the dislike for unknown outsiders. Their reactions were born out of anger and frustration. If there is no cure, why were their sick being taken off to a distant location?

For some villagers, then, these health-care workers were responsible for the disease and they were the ones spreading it. In the village of Kolo Bengou, Guinea, the *New York Times* reported of a 17-year-old Faya Iroundouno leading the campaign of a group of youth against aid groups. "We don't want any

FIGURE 6.5   Acquired from Zen Garden Inspiration Gallery.[38]

FIGURE 6.6   Acquired from *Leadership*, Nigeria's most influential newspaper.[39]

contact with anyone" he said. "Wherever those people have passed, the communities have been hit by illness."[24] At the sight of a Westerner, these villagers would become anxious and would run away, seeing them as the cause of the disease. "Ebola, Ebola!" they would say while running away.[24] In Wabengou, another village in Guinea, the villagers blocked the road to outsiders with a tree. The chief, Marcel Dambadounou, explained, "We don't accept their presence at all. They are the transporters of the virus in these communities"[24] (Figures 6.5 and 6.6). The most terrifying reaction was that of the villagers in Womey, Guinea, in September 2014. A delegation of eight officials and local journalists were sent to the village to warn of the dangers of the Ebola virus disease. They were all killed by a mob and their dismembered bodies were dumped in a septic tank.[25] However, this was not the default reaction of every village.

For a large number of villagers, it did not make sense for their sick to be taken away to a distant hospital. These villagers had seen how most patients, who were taken to the hospitals, would end up dying anyway. A nurse working at the Doctors Without Borders (*Médecins Sans Frontières*) treatment center in Guéckédou succinctly described this sentiment, "Here, if the people come

FIGURE 6.7   A cartoon published in the online Liberian newspaper, *Liberian Observer*.[40]

FIGURE 6.8   A cartoon published in the online *Liberian Observer*.[40]

in, they don't leave alive."[26] A village elder complained about this irony as well, which further explained the apprehension villagers had in sending their sick family members away, "We refused them, because of what was being said; people said that if you send 20 away, 19 would die."[25]

In the major cities, the reaction was slightly different. Many people were angry and a number of them did not want to believe that the Ebola virus disease was real (Figure 6.7). In Liberia, people were skeptical of the so-called "Ebola treatment centers" (Figure 6.8). One center in the capital of Monrovia was attacked by a mob and led to the escape of a number of Ebola virus-infected patients. The mob could be heard angrily saying, "there's no Ebola."[27]

THE UNITED STATES AND EUROPE STANDBY

Eventually, in some corners of society, the initial sentiment of distrust morphed into an overall skepticism with the response of the United States and the rest of the world. *Vanguard*, an online Nigerian newspaper, published an article titled, "Ebola Virus Disease and America's Funny Hypocrisy." In this article, author Rotimi Faan complained about the fact that before the arrival of Ebola virus disease to the United States by way of Thomas Eric Duncan, "for the first three months after the outbreak of the disease, the matter was more or less regarded as the local problem of the affected African countries. Even in Nigeria not many paid any attention to Ebola virus disease until Sawyer berthed in the country in the third week of June and several medical personnel quickly came down with the disease."[15]

Adding to the frustration was the fact that a new drug, ZMapp, had become available in the United States and was being used to treat Ebola virus disease patients there. "It was successfully used for the American," Faan wrote, "but thereafter America would not permit its use elsewhere for different reasons, the major (one) being that it was yet to pass the stage of clinical trial." The same article further stated, "At a time the rest of the world, specifically Africa, reeled with fear of possible annihilation America hoarded its ZMapp and other vaccines that could have alleviated the suffering of thousands of people. They thought nothing of the fact many were dying and many more could die"[15] (Figure 6.9).

According to one cartoonist based in Lisbon, André Carrilho, "People in the African continent are more regarded as an abstract statistic than a patient in the U.S. or Europe."[28] In order to drive the message home, Carrilho, drew a cartoon showing how one white person gets more attention than the large number of black people infected with the Ebola virus (Figure 6.10).

FIGURE 6.9 A sick Africa is calling for help, but the world is asleep. When Africa mentions the epidemic reaching the West, the world wakes up.[41]

FIGURE 6.10 "One Powerful Illustration Shows Exactly What's Wrong with How the West Talks about Ebola" an article in World Mic published this cartoon, by André Carrilho, online.[26]

Adding to this skepticism was the fact that the international community brazenly divided the task of aiding the West African countries based on historical colonial divisions. The United States was responsible for helping Liberia, a nation that was founded in 1822 by freed American slaves; Britain was responsible for its former colony, Sierra Leone; and France was responsible for its former colony, Guinea.[29]

TOO LATE, TOO LITTLE TO BE EFFECTIVE

At the same time, the West African countries began realizing the dimensions of the problem. Different measures were put in place to control the spread of Ebola virus disease. Many education campaigns sprouted up in Guinea, Liberia, Sierra Leone, and Nigeria. In addition, steps were taken by the governments to reign in the epidemic. Across various city and village streets, posters and signs were put up warning against Ebola virus disease and informing people of the necessary precautions one should take (Figures 6.11 and 6.12). In some cases, famous personalities spoke about the epidemic and helped mitigate the "fear of the unknown." Some of these measures were met with the worst resistance. Others paved the way for a better response.

FIGURE 6.11 A street in Monrovia, Liberia, with "Stop Ebola" painted on a wall. The image was published on the *allAfrica* online news Web site.[42]

FIGURE 6.12 Street art depicting the signs and symptoms of Ebola virus disease.[43]

FIGURE 6.13 A youth wounded in clashes lay by the roadside in West Point neighborhood. Soldiers fired live rounds when a crowd tried to break through a barricade.[44] *Image from "Clashes in Liberia After Ebola Quarantine Order" slideshow on the New York Times Web site.*

The Liberian government, in August 2014, began enforcing a mandated quarantine in the West Point neighborhood of the capital city of Monrovia. This step was taken in the hopes of keeping travel to a minimum and allowing health-care workers to move around freely and investigate Ebola virus disease cases and trace contacts. However, like the other problematic aspects of the Liberian epidemic, this was no easy task. People were infuriated as those who relied on travel to other neighborhoods for their livelihoods could no longer work. The *New York Times* article, "Clashes Erupt as Liberia Sets an Ebola Quarantine," quoted a resident complaining, "There is nowhere to go for our daily bread."[30] One resident aptly asked, "You fight Ebola with arms?"[30] Unfortunately, the anger was not limited to furious statements. When some of the angry residents began hurling rocks and trying to storm the barbed wire barricades, soldiers responded with live ammunition rounds. Expectedly, there were injuries. Luckily, there were no reported deaths (Figure 6.13).

On the other hand, some slow but steady measures by the Liberian MOHSW in conjunction with the CDC, WHO, and other organizations did lead to a gradual improvement in diagnosing and reporting of the Ebola virus disease. MOHSW "developed a national task force and technical expert committee to oversee the management of the Ebola-related activities."[31] The CDC further strengthened the response efforts by instituting principles of the incident management system (IMS) that is in place in the United States for, among other things, managing responses to public health emergencies. The IMS consists of defined roles for command, operations, logistics, planning, and finance/administrative functions, and additionally includes the scientific/public health response role.[31] A national strategy, Rapid Isolation and Treatment of Ebola (RITE), was instituted in October, which consisted of the formation of local health teams responsible for the investigation, follow-up, and isolation of Ebola virus disease outbreaks in remote villages.[20] This strategy resulted in some improvement of the disease burden on the nation of Liberia. In the six outbreaks before the start of RITE,

FIGURE 6.14   A page from the info comic, "Ebola Man."[45]

the median time from an outbreak to an alert reaching the local health team was 40 days.[20] For the six outbreaks after the institution of RITE, the median time from an outbreak to an alert reaching the local health team was 25 days.[20]

In Guinea, a famous French soccer player, Lilian Thuram, was visiting the country on a book tour. In an interview with *Jeune Afrique*, he was asked why he was visiting the country when most foreign visitors were avoiding all travel. His response was simple, "On entend effectivement beaucoup de choses dans la presse, mais on sait aussi que cette dernière n'est pas toujours réaliste. Souvenez-vous du traitement médiatique de la crise des banlieues en 2005 par les télévisions américaines... On avait l'impression que la France était en pleine guerre civile! Dans le cas de la Guinée, il faut donc relativiser les choses. Certes, il y a des gens malades, mais ceux-ci sont traités dans les hôpitaux et les modes de transmission de la maladie sont très spécifiques." Roughly translated, Thuram said, "One hears a lot of things in the media but one knows, also, that is not always the reality. You know about the media treatment of the crisis in the suburbs in 2005 by the American television…One had the impression that France was at civil war! The same is the case in Guinea, it is necessary to put things into perspective. Certainly, there are sick people, but they are being treated at hospitals and the modes of transmission of the disease are very specific."[32]

Although Nigeria had a low rate of Ebola virus disease cases, there was a strong campaign to educate people about the dangers of the Ebola virus infection and the modes of transmission. One such campaign was in the form of an eight-page info comic called "Ebola Man" (Figure 6.14).

As is the case with all events that become news-worthy, the pop-culture industry also jumped on the campaign bandwagon. Various singers and actors from a number of West African countries began writing and performing songs about the Ebola virus disease and how it spreads. "Be reassured, the doctor will help you if you have Ebola" is a line in the French-language song, "Africa Stop Ebola."[33] A well-known artist in the region, Peter Cole, put his other projects on hold to release a song titled "Ebola is Real." "Don't Deny it; Don't Defy It; It's a danger for your community," he sings in the hopes of encouraging people to bravely come to terms with the reality of the disease. Cole's song came at an optimal time, as Senegal had just reported their first case in August 2014. Cole was also looking into doing a video of the single, "I hope to find footage of Ebola management and have it in a video together with the song."[34]

We might never know whether these campaigns alone succeeded in convincing people of the existence of the Ebola virus disease. We might also never know if it was solely these campaigns that convinced some villages to open their doors.

OUR DOORS ARE OPEN; COME TAKE OUR SICK

There most certainly was a noticeable correlation between the increased campaigns and village chiefs opening their doors to health-care workers. In Dandano, Guinea, that is exactly what happened. A November 2014 *New York Times* article reported on the arrival of a Red Cross team and the interaction with the villagers. "'Bring out your sick!' the chief shouted angrily at the crowd, shaking his fist and warning of illness and death for the whole village if it did not obey. 'Don't hide them!' he yelled. 'If you don't expose them, you will suffer!'"[25] This was a big relief, as Dandano is located in the district of Macenta, which, as of November 2014, was the worst-hit region in Guinea. One can only imagine how much courage it must have taken for village elders to open their doors to a perceived—albeit incorrectly—enemy and transmitter of the deadly disease. Even more courageous was their willingness to publically accept that their initial reaction was flawed: "We were wrong. But we didn't understand the cause of this sickness."[25]

In some parts of West Africa, the openness of the villagers may be attributed to some success stories of patients in treatment centers. In the city of Bo, Sierra Leone, Doctors Without Borders (*Médecins Sans Frontières*) reported of the successful management of three Ebola virus disease patients, who they referred to as "three miracles."[35] One of the patients, given the pseudonym Hassan, was in grave condition and was described as "almost dead."[35] He was in a state of confusion, lethargy, and disorientation, and he had "constant diarrhea."[35] The nurses dedicated half an hour at a time to feed him and give him water. As a result, one day he was talking again and told his doctor that he would start walking tomorrow. Sure enough, the next day he began walking again.[35]

His friend, given the pseudonym Mohamed, was also admitted to the treatment center at Bo, Sierra Leone. One night, he became angry and confused and began threatening other patients. Monica Arend-Trujillo, the doctor chronicling the recovery of these patients described the fear that is associated with this state, "It's usually a very bad sign when people develop mental problems."[35] That night, an antipsychotic drug had to be administered to calm Mohamed down. When he woke up, he was calm again and recovered well enough to start playing cards with Hassan and other recovering patients. In a few days, both Hassan and Mohamed were discharged, free of the Ebola virus disease.

Admitted with Hassan was his 10-month-old niece and her Ebola virus-infected mother. The niece was being breastfed and had arrived with her mother and other Ebola virus-infected patients in an ambulance. Despite this, the child survived; she was placed in a separate cubicle and when she developed fever she was given antibiotics. Fortunately, her Ebola virus detection test returned negative and she was cleared for discharge.[35]

In addition to the work of Doctors Without Borders/Médecins Sans Frontières, the West African nations received assistance from the CDC, and the United States Agency for International Development (USAID). USAID activated a Disaster Assistance Response Team (DART) with members in Monrovia, Liberia and Conakry, Guinea.[36] The public health and medical response positions on the DART were staffed by the CDC.[36] In Monrovia, the United States began building a 25-bed critical care hospital with capabilities to manage Ebola virus disease patients. This hospital was to be staffed with US-licensed medical professionals.[36] A CDC study showed that in Liberia, the CDC's work in creating an IMS and sending out health workers to remote and inaccessible areas greatly improved the situation, although the effect was noted after the Ebola virus disease had already caused much damage.[20]

The most unique response of outside nations was the deployment of troops. The United States deployed 2400 troops to Liberia to build treatment centers, establish mobile blood laboratories, and transport Ebola virus treatment supplies around Liberia.[29] This had led people to ask if the United States would consider sending troops to other affected nations such as Sierra Leone. Britain deployed close to 800 soldiers in Sierra Leone and according to British officials they were building new treatment centers and training medics.

Despite these positive signs, the epidemic continues to ravage West Africa and people have become weary of the effectiveness of these measures. Those who have contracted the disease still have no "magic," "cure-all" drug and their family members do not know what to expect. In the midst of this, we are sure to see continued confusion and a myriad of repercussions, including economic and political consequences.

REFERENCES

1. Organization, W.H. *Ground zero in Guinea: the outbreak smoulders—undetected—for more than 3 months*. 2013. Available from: http://www.who.int/csr/disease/ebola/ebola-6-months/guinea/en/.
2. Baize S, Pannetier D, Oestereich L, Rieger T, Koivogui L, Magassouba N, et al. Emergence of Zaire Ebola virus disease in Guinea. *N Engl J Med* October 9, 2014;**371**(15):1418–25.
3. Beukes S. *Ebola: finding patient zero*. October 27, 2014. Available from: http://blogs.unicef.org/2014/10/27/ebola-finding-patient-zero/.
4. Yan H, Smith E. *Ebola: who is patient zero? Disease traced back to 2-year-old in Guinea*. October 28, 2014. Available from: http://www.cnn.com/2014/10/28/health/ebola-patient-zero.
5. Organization, W.H. *Ebola virus disease in Guinea*. March 23, 2014. Available from: http://www.afro.who.int/en/clusters-a-programmes/dpc/epidemic-a-pandemic-alert-and-response/outbreak-news/4063-ebola-hemorrhagic-fever-in-guinea.html.

6. Ebola virus disease in West Africa—the first 9 months of the epidemic and forward projections. *N Engl J Med* 2014;**371**(16):1481–95.

7. Dixon MG, et al. Ebola viral disease outbreak—West Africa, 2014. *MMWR Morb Mortal Wkly Rep* 2014;**63**(25):548–51.

8. Fink S, Grady D. *Tracing Ebola's breakout to an African 2-Year-Old*. August 9, 2014. Available from: http://www.nytimes.com/2014/08/10/world/africa/tracing-ebolas-breakout-to-an-african-2-year-old.html?_r=0.

9. Ross W. *Ebola crisis: how Nigeria's Dr Adadevoh fought the virus*. October 20, 2014. Available from: http://www.bbc.com/news/world-africa-29696011.

10. Cohn J. *The heroic, tragic story of the doctor who saved Nigeria from Ebola*. October 25, 2014. Available from: http://www.newrepublic.com/article/119956/doctor-stella-adadevoh-isolated-ebola-case-stopped-nigeria-outbreak.

11. Soleye N. *Remembering my aunt, Dr. Ameyo Adadevoh, who stopped Ebola in Nigeria*. October 09, 2014. cited 2014; Available from: http://www.msh.org/blog/2014/10/09/remembering-my-aunt-dr-ameyo-adadevoh-who-stopped-ebola-in-nigeria-0.

12. Ogunlesi T. *Dr Stella Ameyo Adadevoh: Ebola victim and everyday hero*. October 20, 2014. Available from: http://www.theguardian.com/lifeandstyle/womens-blog/2014/oct/20/dr-stella-ameyo-adadevoh-ebola-doctor-nigeria-hero.

13. allAfrica. *Liberia: Patrick Sawyer lied about 'Ebola' contact, late doctor's note reveals*. August 25, 2014. Available from: http://allafrica.com/stories/201408261017.html.

14. Wilson J. *Ebola fears hit close to home*. July 30, 2014. Available from: http://www.cnn.com/2014/07/29/health/ebola-outbreak-american-dies/.

15. hypocrisy, E.V.D.a.A.s.f.. *Ebola virus disease and America's funny hypocrisy*. October 08, 2014. Available from: http://www.vanguardngr.com/2014/10/ebola-virus-disease-americas-funny-hypocrisy/.

16. Oleribe OO, et al. Ebola virus disease epidemic in West Africa: lessons learned and issues arising from West African countries. *Clin Med* 2015;**15**(1):54–7.

17. npr. *Mali's first Ebola case in current outbreak is 2-year-old girl*. October 24, 2014. Available from: http://www.npr.org/blogs/thetwo-way/2014/10/24/358517723/malis-first-ebola-case-in-current-outbreak-is-2-year-old-girl.

18. McNeil Jr DG. *Second Ebola outbreak in Mali Eclipses early success*. November 12, 2014. Available from: http://www.nytimes.com/2014/11/13/health/mali-reports-a-second-larger-ebola-outbreak.html?_r=1.

19. Organization, W.H. *Mali case, Ebola imported from Guinea*. November 10, 2014. Available from: http://www.who.int/mediacentre/news/ebola/10-november-2014-mali/en/.

20. Kateh F, et al. Rapid response to ebola outbreaks in remote areas—Liberia, July–November 2014. *MMWR Morb Mortal Wkly Rep* 2015;**64**(7):188–92.

21. Butler YS. *Ebola virus*. November 5, 2014. Available from: http://www.mayoclinicproceedings.org/article/S0025-6196%2814%2900882-9/fulltext.

22. Blackley DJ, et al. Rapid intervention to reduce ebola transmission in a remote village—Gbarpolu county, Liberia, 2014. *MMWR Morb Mortal Wkly Rep* 2015;**64**(7):175–8.

23. Incident Management System Ebola Epidemiology Team, C.D.C., et al. Update: Ebola virus disease epidemic—West Africa, February 2015. *MMWR Morb Mortal Wkly Rep* 2015;**64**(7):186–7.

24. Nossiter A. *Fear of Ebola breeds a terror of physicians*. July 27, 2014. Available from: http://www.nytimes.com/2014/07/28/world/africa/ebola-epidemic-west-africa-guinea.html?_r=0.

25. Nossiter A. *Fear of Ebola opens wary villages to outsiders in Guinea*. November 16, 2014. Available from: http://www.nytimes.com/2014/11/17/world/africa/fear-of-ebola-opens-wary-villages-to-outsiders-in-guinea.html?_r=0.

26. News, T.N.Y. *Fear of Ebola breeds a terror of physicians.* July 27, 2014. Available from: http://www.nytimes.com/2014/07/28/world/africa/ebola-epidemic-west-africa-guinea.html?_r=0.

27. Africa, N. *Ebola crisis: confusion as patients vanish in Liberia.* August 17, 2014. Available from: http://www.bbc.com/news/world-africa-28827091.

28. Kleeman S. *One powerful illustration shows exactly what's wrong with how the west talks about Ebola.* October 7, 2014. Available from: http://mic.com/articles/100618/one-powerful-illustration-shows-exactly-what-s-wrong-with-media-coverage-of-ebola.

29. Gettleman J. *As Ebola rages, poor planning thwarts efforts.* December 6, 2014. Available from: http://www.nytimes.com/2014/12/07/world/africa/as-ebola-rages-in-sierra-leone-poor-planning-thwarts-efforts.html.

30. Onishi N. *Clashes erupt as Liberia sets an Ebola quarantine.* August 20, 2014. Available from: http://www.nytimes.com/2014/08/21/world/africa/ebola-outbreak-liberia-quarantine.html?_r=0.

31. Pillai SK, et al. Developing an incident management system to support Ebola response—Liberia, July–August 2014. *MMWR Morb Mortal Wkly Rep* 2014;**63**(41):930–3.

32. racisme(541). *Lilian Thuram: "Il n'y a pas de sujet tabou, on peut et on doit parler de tout".* April 30, 2014. Available from: http://www.jeuneafrique.com/Article/ARTJAWEB20140430180937/dition-racisme-lilian-thuram-homophobie-tol-rance-lilian-thuram-il-n-y-a-pas-de-sujet-tabou-on-peut-et-on-doit-parler-de-tout.html.

33. Ebola, A.S. *Africa stop Ebola.* October 27, 2014. Available from: http://allafrica.com/view/resource/main/main/id/00090815.html.

34. allAfrica. *Liberia: doing it for Ebola—Peter Cole unleashes awareness song.* September 5, 2014. Available from: http://allafrica.com/stories/201409051731.html.

35. Arend-Trujillo DM. *Ebola: "Three miracles" in Bo, Sierra Leone.* November 20, 2014. Available from: http://www.doctorswithoutborders.org/article/ebola-three-miracles-bo-sierra-leone.

36. USAID. *Ebola.* 2014. Available from: http://www.usaid.gov/ebola.

37. World, C. *The Ebola epidemic.* 2014. Available from: http://www.cnn.com/2014/04/04/world/gallery/ebola-in-west-africa/index.html?iid=article_sidebar.

38. Gallery, Z.G.I. *U.S. attempts to fight Ebola, Ukraine ratifies landmark deal Amid Russian sanctions and U.S. steps up plans to fight ISIS as al-Qaeda expands.* September 18, 2014. Available from: https://craftymcclever.wordpress.com/tag/ebola-u-s-response/.

39. Ishiekwene A. *We got it right on this one.* August 29, 2014. Available from: http://leadership.ng/columns/382343/got-right-one.

40. Lumeh AL. December 11, 2014; Available from: http://www.liberianobserver.com/cartoon/cartoon-october-31-2014.

41. peace, d. 2014; Available from: http://desertpeace.files.wordpress.com/2014/08/ebola-english.gif.

42. South African Government. *Africa: G20 calls for unity in Ebola fight.* November 15, 2014. Available from: http://allafrica.com/view/group/main/main/id/00033922.html.

43. rollcall. 2014; Available from: http://blogs.rollcall.com/white-house/wp-content/uploads/sites/16/2014/09/454957958.jpg.

44. Times, N.y. 2014; Available from: http://www.nytimes.com/2014/08/21/world/africa/ebola-outbreak-liberia-quarantine.html?_r=0#slideshow/100000003065263/100000003065357.

45. Akatasia. Ebola Man, an info comic bringing awareness to the Ebola Virus Disease (EVD). September 17, 2014. Available from: http://akatasia.com/articles/ebola-man-info-comic-bringing-awareness/.

Chapter 7

Ebola Virus Disease in the United States

Chapter Outline

Ebola virus disease, a rare, viral entity only known for outbreaks localized to remote parts of Africa, for the first time in history, has reached the United States soil and is now a topic of commotion and chaos, wreaking havoc in the minds of the United States citizens. Although a rare complication, the virus is best known for making an individual bleed from all orifices of the body and killing victim it infects. All major news stations have covered the ill-fated germ. Ebola appears to have infected the media, causing uproar of concern and paranoia among the population in the United States.

JULY 2014: FIRST RESPONSE TO AFRICAN OUTBREAKS

The Centers for Disease Control (CDC) issues a level 3 travel warning to travelers planning to go to the West African nations of Guinea, Sierra Leone, and Liberia, meaning that one should avoid all nonessential travel to these regions. This is a subtle indication of the worsening Ebola outbreak in Africa. Fifty members of the CDC are going to make their way to affected regions to help with the outbreak. Although there is some awareness, the Ebola virus disease still is not causing a media bonanza that proved to happen at a later point.

AUGUST 2014: EBOLA VIRUS STRIKES
A UNITED STATES CITIZEN

The CDC issues a level 2 travel warning to Nigeria, due to a spread of infection to the neighboring country. A single case of Ebola virus disease is reported in Senegal, yet another indicator of the disease is spreading like wildfire. Over 50 CDC staff members are now in West Africa assisting with the Ebola virus disease outbreak. Their jobs include training and coordinating health-care workers, as well as collecting and analyzing data. CDC laboratory technicians train lab workers to create laboratories for Ebola virus disease testing. Some CDC workers serve to educate locals on the prevention and importance of getting medical care early if they develop symptoms of Ebola virus disease. Other CDC health professionals work with border and airport employees in West Africa. The CDC has also posted infection prevention and control guidelines for the United States hospitals in the case of an Ebola virus disease patient walks through their doors.

The real uproar seemed to have started when the United States missionary, Dr Kent Brantly, 33 years old, was flown out of Liberia on August 2, 2014. His arrival at Emory University Hospital was shortly followed by US aid worker 59-year-old Nancy Writebol on August 5, who was also servicing in Liberia. Although these US aid workers do not remember the specific event that led to them contracting the virus, they remember points in time where it could have been possible. A 3-week stay at Emory, world-class care, and experimental Ebola virus infection therapeutics all seemed to contribute to their recovery after 3 weeks of admission. These were the first two cases of Ebola virus disease that were treated on the United States soil. Although their care on the United States grounds caused concern of spread to other individuals, their treatments and isolation proved successful, with the patients fully recovered and no others infected as a result of their admission.[1,2]

The first patient, Dr Kent Brantly, being a father of two and husband of Amber Brantly, is considered a loving family man. Although Ebola virus disease caused him much pain, he believed his mission trip to Liberia was ordained by God. Brantly has always been a helping hand since he was young. He worked as a family physician in Texas, helping people day by day, and even directed the group's Ebola consolidated Case Management Center in Liberia. He has even donated his serum to a number of Ebola-infected patients in the United States. He is the type of person who sees the glass half full rather than half empty, believing that his experience with Ebola was not a tragedy, but a blessing.[1,2]

SEPTEMBER 2014: THE NUMBER OF PATIENTS
IN THE UNITED STATES INCREASES

The CDC appears to have realized that the Ebola virus disease is successfully outpacing their efforts. It is clear that despite the help provided at this point, there is a severe lack of resources and funding. The window of opportunity to help halt the outbreak continues to narrow despite responses from the United States and CDC. On the bright side, it appears that the caseload in Nigeria and Senegal appears stable, despite the exponentially rising cases in Guinea, Sierra Leone, and Liberia.

A third US aid worker, Dr Rick Sacra, aged 51, was admitted to the University of Nebraska Biocontainment Unit on September 5.[3] He remembers doing many cesarean sections, working as an obstetrician in Liberia and could have contracted the virus during the procedures. There are no other details as to how he may have contracted the virus. After nearly 3 weeks of intensive care, two serum transfusions from Dr Kent Brantly, and experimental Ebola therapeutics, Dr Rick Sacra was discharged from the University of Nebraska, his blood virus-free. A fourth US aid worker, and the third patient to be treated at Emory University, was admitted on September 9 and has remained under the radar as far as identification. All that is known is that this patient has the longest known admission for Ebola virus disease and was discharged on October 19, virus-free.

Sacra, having to overcome the tremendous virus, did not talk much about himself during interviews. Being a very thoughtful and caring man, Sacra was worried about the others who have been affected by Ebola and how they were feeling. Even though he contracted his infection in Liberia, he hopes to go back and improve the country's health system. Sacra says he is grateful that he was able to benefit from Brantly's serum transfusions and is not sure whether he would have been able to recover without them.[3]

To be completely clear, the four aforementioned cases were all diagnosed in Africa, and the patients were then transferred to the United States for treatment. These patients, in their fight with Ebola, have not passed the infection to any other persons in the United States. These cases made history, as they were the first cases of Ebola virus disease diagnosed in Africa to be treated in the United States.

On September 25, a man by the name of Thomas Eric Duncan, aged 42 years, was admitted to Texas Presbyterian Hospital, with the symptoms of abdominal pain and fever of 103°F, which began on the night of the 24. He was discharged home that same day on antibiotics, despite telling the nurse he recently had traveled to Liberia. He was diagnosed with a "common viral disease," and sent home on antibiotics, despite antibiotics having no action against viral illnesses. He began vomiting on September 28 and was readmitted again to the same hospital. During this admission, he was suspected to have Ebola virus disease and was confirmed to have tested positive on September 30. Thomas was the first ever case of Ebola virus disease to be diagnosed on the United States grounds. Duncan came to the United States to visit family and friends, specifically his son and the mother of his son. Duncan started showing symptoms 4–5 days after the trip. Sadly, Duncan was diagnosed too late, slimming his chance of survival, and passed away on October 8, 2014.[4]

The CDC knows that this single case is a cause of major concern, as now the possibility of spread to the United States has become a reality. The CDC's anticipation and preparation for this first case included implementing surveillance

and laboratory methods in each state to detect cases, creating tools and protocols for health departments to conduct public health investigations, providing
recommendations for infection control among other measures for prevention
of disease spread, guiding airports and flight crews in the event of a suspected
case, creating emergency medical services at airports, asking border protection
and customs officers about reporting suspected cases to the CDC, and making
up-to-date information available to the public, travelers to international areas, as
well as public health partners.

OCTOBER 2014: GAPS IN THE UNITED STATES HEALTH SYSTEM

Duncan was thought to have contracted Ebola from a pregnant woman in Liberia,
whom he had helped into a vehicle. It was reported that he was unaware she had
Ebola virus disease and thought her difficulties were related to her pregnancy.
Duncan's condition quickly deteriorated. Experimental therapeutics was not
employed until several days after his admission, when he was already intubated
and in critical condition. He was offered blood from the Ebola virus disease
survivor, Kent Brantly, but unfortunately was not a biological match. He passed
away on October 8. Questions arose as to whether his care was suboptimal.
Duncan was the first ever case of Ebola diagnosed in the United States and the
first ever death from Ebola in the United States.[4]

Duncan's case prompted major concern across the country, such that the
CDC implemented arrival screening at five major US airports, as these airports
receive about 94% of travelers from affected countries. Screening includes travelers from Guinea, Liberia, and Sierra Leone to be escorted to an area of the
airport set aside for screening where trained airports staff observe for signs of
illness, ask incomers a series of questions, provide information regarding Ebola,
and give reminders to look out for symptoms. Temperature of these travelers
will be taken with a noncontact thermometer. If a fever or symptoms are present
or the health questionnaire points to possible exposure, a CDC quarantine-sta-
tion public health officer will evaluate them. The officer will repeat a temperature reading and make a public health assessment. Travelers who are deemed to
require further evaluation or monitoring will be taken care of by the appropriate
public health authority. Travelers from these countries who have no indications
of exposure will receive health information on Ebola.[4]

Two days before Duncan's death on October 6, an NBC news cameraman,
Ashoka Mupko, was the second person admitted to Nebraska medical center,
and the fifth person to contract Ebola in Africa and be flown to the United States
for treatment. He remembers being around many sick people the week before he
became symptomatic, but cannot remember a specific moment where he could
have contracted the virus. He was declared Ebola-free on October 22.[5]

Duncan's case remained controversial, especially regarding possible suboptimal treatment. Something that definitely proved to be suboptimal was the
biohazard protocol that was employed at the Texas Presbyterian Hospital, as two

nurses that cared for Duncan were confirmed to have Ebola virus disease. Their names are Nina Pham and Amber Joy Vinson.

Nina Pham, a 26-year-old nurse, was the second case of Ebola virus disease diagnosed in the United States, and the first case of transmitted Ebola virus disease in the United States. Nina Pham spiked a fever on October 10 and subsequently admitted herself into Texas Presbyterian Hospital. She was diagnosed with Ebola virus disease on October 12 and was transferred to the National Institute of Health on October 14 for further treatment. She received a plasma transfusion from Kent Brantly. She does not remember an event where she could have contracted the virus. On October 24, Nina was discharged from the National Institute of Health, virus-free.[6]

Pham, fully recovered, was very excited to see her best friend, a cocker spaniel named Bentley after his quarantine shortly after her diagnosis. He is doing well and not displaying any signs of Ebola. His stool tested negative for Ebola once and was tested again before his release from quarantine on November 1.[7] "Bentley re-entering my life is another reminder of hope and encouragement for me moving forward and living my life to its fullest with my best friend at my side again," said Pham.[7]

Amber Joy Vinson, a 29-year-old nurse, is the second case of Ebola virus disease through transmission on US soil. She was also one of the nurses to have treated Duncan. She reported a temperature the morning on October 14 of 100.3°F to the Department of Health, and then admitted herself to Texas Presbyterian Hospital. She was on a flight returning from Dallas 1 day prior to reporting a fever. She was transferred to Emory University Hospital on October 15, becoming their fourth Ebola admission. On October 23, Amber was declared virus-free and was discharged on October 28, 2014.[8]

Before Vinson contracted Ebola virus disease, she was preparing for her wedding. Instead of planning for her big day, she had to stay in isolation.[8] Many family members and friends consider Vinson to be sweet and kind, telling the press that Vinson was drawn to the idea of becoming a doctor since she was young, and has always been considered generous and compassionate. "She's a very good person who did very good work and deserves to be honored, not detained, for it," said Hyman, her lawyer. Vinson is now happy and is working to becoming healthier and stronger after Ebola struck her system.[8]

On Monday, October 20, the CDC tightened the protocol for care of infected patients. This appears to have been a result from controversy in light of the aforementioned cases of Ebola being transmitted to the two nurses who cared from Duncan. Hospitals have begun implementing the new guidelines. The new guidelines include using personal protective equipment (PPE) that covers every part of the health-care provider's skin. Equipment prior to this kept areas, such as the neck, significantly exposed. Another major change is that the use of PPE be monitored, along with designated areas for putting on and taking off PPE. The CDC vows to send Ebola response teams to any hospital in the United States with a confirmed case.

The CDC has decided to step up airport checks for people returning to the United States. Although there are temperature checks when departing from a West African country, the United States health authorities have decided to also perform temperature checks upon arriving at airports across the country. The five major airports implementing Ebola screening upon entry are at New York's John F. Kennedy International Airport, New Jersey's Newark Liberty Airport, Virginia's Washington Dulles Airport, Chicago O'Hare Airport, and Atlanta's Hartsfield Jackson Airport. These are the designated airports because around 95% of those returning from West Africa will pass through these locations. These airports will implement a few basic layers to this approach. Each person arriving to will receive health information about Ebola virus disease, indicating the symptoms to look out for. They will also have to answer questions, have their temperature taken, and be observed for signs of Ebola. In the event that they are deemed to possibly be suffering from Ebola virus disease, they will be isolated and taken to a hospital for evaluation, testing, and treatment.

NOVEMBER 2014: EBOLA VIRUS DISEASE REACHES ITS FIRST PEAK

Craig Spencer is the second case of travel-associated Ebola virus disease in the United States, second to Duncan. He is a 33-year-old American doctor who works at New York Presbyterian Hospital. He was working in Guinea and was a Missionary for Doctors Without Borders (*Medecins Sans Frontieres*). He arrived back to the United States at John F. Kennedy Airport on October 17. On October 23, he reported a fever. He was confirmed to be positive for the deadly virus. From the time of his arrival in New York, until the day he spiked a fever, he was in contact with only a few people, and had done things like get a ride in an Uber cab service vehicle, and gone bowling. Many people criticize him for going out into the public and enjoying his time back, but the facts show that the virus can only be transmitted once a patient is symptomatic, which Spencer was not.

Spencer was treated at Bellevue Hospital, was discharged on Tuesday, November 11, 2014, and is now Ebola-free. A press release statement from the Doctors Without Borders (*Medecins Sans Frontieres*) Web site was released by Dr Craig Spencer. He outlines his thanks to the Bellevue medical team, especially to his attending physician, Dr Laura Evans. He states that, although his treatment was exceptional at Bellevue Hospital and has had endless media attention, it is important to remember the real problem lies in West Africa. Efforts and attention should be focused to those who are in the real heat of danger.[9]

Along with Spencer being quarantined, his fiancé Morgan Dixon, age 30, was quarantined. Both are doing great now and are preparing for their wedding, which will be happening fall 2015. Since Spencer's ordeal occurred in the New York City, New Jersey Governor, Chris Christie, wrongly forced isolation on Kaci Hickox, a nurse who treated Ebola victims in West Africa. This is yet another sign of Ebola paranoia growing as political figures set rules on Ebola despite having no background in the medical field or using scientific facts to justify the cause. It could be argued that the Governor does not have the ability to make clear decisions and is trying to win over his community by feeding the fire of their fear, a questionable stunt for votes. A health-care worker who goes in the line of duty to help people intuitively has the ability to be cautious with their own health and, more importantly, to keep the infection from spreading.[9]

In the midst of her time in New Jersey, Dixon decided to head to Maine to live with her boyfriend. She was welcomed to Maine with more ignorance from Governor Paul LePage, who ordered her to be quarantined despite testing negative for the virus. Despite his decision, she defied his orders, reminding the nation of her civil liberties. Governor LePage's nonsense was overturned when a judge ordered that her restrictions were too stringent and to only continue her regular monitoring.[10]

Forced quarantine can be discouraging for health-care workers to volunteer in West Africa, where the source of the issue is. Without fighting this battle at the source, it will only continue to ravage and grow. It is frightening to realize that there are political leaders in the United States who do not realize the consequence of their forced quarantine. Their actions may have influenced many volunteers to back out of the fight. In essence, some people, who could have been saved in West Africa, may have died because of a lack of volunteers. Political figures must think before they speak because their actions may have cost the lives of others.

The second patient to die from Ebola virus disease in the United States happened in November 16, 2014. Dr Martin Salia, aged 44, contracted the virus while in Sierra Leone. He worked as a general surgeon at Kissy United Methodist Hospital. It is believed that he did not contract the Ebola virus disease from the hospital he worked in, but in other facilities where he volunteered. After landing in the United States on November 15, 2014, Salia received treatment in Nebraska Medical Center's Biocontainment Unit. Doctors treating Salia claimed that he was already in a very critical condition when he arrived to the center. Salia was given the experimental drug ZMapp, a drug given to two Ebola survivors. He also received a serum transfusion from an Ebola virus disease survivor, which was supposed to help his immune system fight this battle. Sadly, Salia passed away on November 16, 2014, and will be missed by his family members, especially his wife, Isatu Salia and his children. On the same day, the CDC added Mali to the list of Ebola-affected areas for which monitoring and enhanced screening methods will be implemented upon entry into the United States.[11]

THE NATIONAL RESPONSE TEAMS STEP UP EFFORTS FOR RECOGNITION AND PREVENTION

Given the pressure of increasing number of cases in the United States, the CDC has implemented many new methods for Ebola preparedness and created a dedicated Ebola team that will attend to any location where there is a patient who requires treatment. Online learning modules have been offered to hospital across the nation. Caregivers identified as leaders for Ebola treatment have taken 3-h classes that provide hands-on training. In some hospitals, the electronic medical records system had been updated to contain tools that help providers screen and triage Ebola virus disease patients. An example of instructions given by the learning module is to place a banner on patient charts notifying others that a patient has risk factors for Ebola virus disease.

The measures for PPE have also been further updated by the CDC. It now includes two tiers of protection. Tier one being less protective, for setting like clinics and ambulatory care. Tier two being for patients exhibiting symptoms of vomiting and diarrhea where risk of transmission through bodily fluid is likely. Many hospitals have had their leading infectious disease experts give grand rounds, which is the largest educational lecture that a hospital can provide to its staff, on Ebola virus disease in an effort to further educate its health-care providers.

Despite all the Ebola events in the United States, only a fifth of the nation that was randomly polled in October seems to be actually concerned about Ebola. As far as trusting the United States with Ebola preparedness, Pam Neary, aged 59 years, a former Afton Democrat who served the Minnesota Legislature from 1993 to 1994 says "This is like the fear of flying to me, You have to say to yourself, I don't think this pilot wants to go down any more than I do, so I'm just going to have to trust him."[12]

REFERENCES

1. CBS News. *"Miraculous day": American Ebola patients discharged from Atlanta hospital.* 2014 [cited 10.12.2014]. Available from: http://www.cbsnews.com/news/ebola-patients-kent-brantly-and-nancy-writebol-discharged-from-hospital/.
2. *Exclusive: Ebola survivor Dr Kent Brantly reveals near-death ordeal.* NBC News; 2014 [cited 10.12.2014]. Available from:. http://www.nbcnews.com/storyline/ebola-virus-outbreak/exclusive-ebola-survivor-dr-kent-brantly-reveals-near-death-ordeal-n194111.
3. Fox M. *Ebola patient Dr Rick Sacra gets 'everything we had' NBC news.* NBC News; 2014 [cited 10.12.2014]. Available from: http://www.nbcnews.com/storyline/ebola-virus-outbreak/ebola-patient-dr-rick-sacra-gets-everything-we-had-n201426.
4. Almendrala A. *What we know about Thomas Eric Duncan, the first Ebola patient diagnosed in the U.S.* The Huffington Post; 2014 [cited 10.12.2014]. Available from: http://www.huffingtonpost.com/2014/10/06/thomas-eric-duncan-dallas-ebola-patient-us_n_5942150.html.
5. McDermott J. *U.S. Journalist Ashoka Mukpo says his body was 'at war' with Ebola.* 2014 [cited 10.12.2014]. Available from: http://www.huffingtonpost.com/2014/10/25/ashoka-mukpo-ebola_n_6048248.html.
6. Yan H. *More than a nurse: who is Ebola patient Nina Pham?* CNN; 2014 [cited 10.12.2014]. Available from: http://www.cnn.com/2014/10/14/health/ebola-who-is-nina-pham/.
7. Ahmed S. *Dallas nurse Nina Pham reunited with her dog, Bentley, after Ebola quarantine.* CNN; 2014 [cited 10.12.2014]. Available from: Dallas nurse Nina Pham reunited with her dog, Bentley, after Ebola quarantine.

8. BLINDER A. *Amber Joy Vinson, Dallas nurse treated for Ebola, is released from hospital. The New York Times*; 2014 [cited 10.12.2014]. Available from: http://www.nytimes.com/2014/10/29/us/ebola-outbreak-dallas-nurse-amber-joy-vinson.html?_r=0.
9. Rainone C. *Who is Craig Spencer, 1st New York Ebola patient?* NBC; 2014 [cited 10.12.2014]. Available from: http://www.nbcbayarea.com/news/national-international/Doctor-Craig-Spencer-New-York-Ebola-Patient-Africa-280273962.html.
10. Hickox K. *Nurse who fought Ebola quarantine, to leave Maine: report. The Huffington Post*; 2014. Available from: http://www.huffingtonpost.com/2014/11/08/kacii-hickox-leaving-maine_n_6127146.html.
11. Roels TH, Bloom AS, Buffington J, Muhungu GL, Mac Kenzie WR, Khan AS, et al. Ebola hemorrhagic fever, Kikwit, Democratic Republic of the Congo, 1995: risk factors for patients without a reported exposure. *J Infect Dis* February 1999;**179**(Suppl. 1):S92–7. PubMed PMID: 9988170. Epub 1999/02/13. eng.
12. Olson J. *Minnesota poll: politics shape Ebola fears.* 2014 [cited 16.12.2014]. Available from: http://www.startribune.com/lifestyle/health/280473152.html.

Chapter 8

Structure and Genetics of Ebola Virus Disease

Chapter Outline

In order to understand how the Ebola virus disease can be treated and what can be done to invent or discover treatment options, it is important to have a clear understanding what we are fighting against, what Ebola virus is made of, how it survives, and how it replicates in human body.

EVOLUTION IN OUR UNDERSTANDING OF VIRUSES

In 1892, Dmitri Ivanovsky studied tobacco mosaic virus disease. He noticed that the agents that caused this disease were so small in size that they passed through porcelain filters that stopped bacteria. In 1898, Marcus Beijerinck made the identical observation and proposed that the agent causing this disease must be a new creature unknown to the world yet different from bacteria and fungi.[1] Investigators were clear about one fact: viruses are small in size and can pass through porcelain filters. The virus got its name from Latin word virulentus meaning "*slimy liquid*" or "*poison.*"

Viruses are not animals, plants, or bacteria and are not living organisms in the literal sense of the word. Viruses are dependent on their host to carry out their life-sustaining functions and reproduce. They cannot manufacture proteins because they lack ribosomes and must use ribosomes from the host cells to produce their own proteins with the help of viral messenger ribonucleic acid (RNA). Ribosomes are present within the cells and are involved in protein formation. Viruses cannot generate energy but derive it from host cells in the form of adenosine triphosphate. These energy molecules are then used to form building materials such as amino acids, nucleotides, and lipids (fat).

In the1930s, ultracentrifuge was developed in Sweden by Theodor Svedberg.[2] E.G. Pickels and J.W. Beams in the United States made further advancements in the practical use of ultracentrifuge techniques.[3] The ultracentrifuge was a revolution in two important ways; it not only enabled scientists to perform physical analysis of proteins, but also made it possible to purify almost any particulate material, protein, or virus by spinning at high speed. With the help of centrifugation techniques, scientists were able to calculate molecular weights of viral particles, which led them to the conclusion that virus particles are objects that may contain large protein molecules.

In 1935, Wendell Stanley in the United States further discovered and published the first proof of the tobacco mosaic virus causing tobaccos's mosaic disease which could be crystallized. He also reported the small amount of phosphorous contamination within the protein "crystals." Using physical techniques, he found that protein of tobacco mosaic virus composed of large regular particles and had a very high molecular weight.[4] Norman Pirie and Frederick Bawden challenged the composition of tobacco mosaic virus being made up of only protein. They showed that RNA, which consists of ribose sugar molecules linked by phosphate groups, could be isolated from tobacco mosaic virus cells as well as other plant viruses, accounting for phosphorus "contamination." This discovery led to the realization that tobacco mosaic virus and other particles of plant virus (virion) were actually the proteins linked with nucleic acid or nucleoproteins (NPs).[5]

In 1939, Gustav Kausche, Edgar Pfankuch, and Helmut Ruska made the first electron microscopic image of tobacco mosaic virus. Helmut was the brother of Nobel Prize winner, Ernst Ruska, who developed the electron microscope. Helmut was the one who first imaged the virus particles by deflecting off the beam of electron over the virus particles coated in heavy metal atoms. On June 7,1908, Helmut Ruska was born in Heidelberg, the sixth of seven children of the historian of science Julius Ruska (1867–1949). He graduated as a doctor in 1932 after studying medicine in Berlin, Innsbruck, and Heidelberg where he prepared a biochemical thesis under the supervision of Ludolf von Krehl (1861–1937). From 1933 to 1940, he served as a doctor in several hospitals in Heidelberg and Berlin. In Berlin, he began to focus on the applications of electron microscopy, particularly on the biomedical problems. From 1938 through the early 1940s, he imaged and showed that virions consist of regular and sometimes complex particles. He imaged virions of poxviruses, tobacco mosaic virus, varicella-zoster herpes virus, and bacteriophages. In 1940 while working in the Siemens & Halske Laboratory, H. Ruska, Pfankuch and Kausche published two short papers showing the first pictures of bacteriophages in the world literature.[6] Ruska showed that bacteria were surrounded by large numbers of round particles in broth culture, where bacteria were mixed with a phage suspension at a magnification of 14,000 times. These images caused considerable excitement in Europe. At a magnification of 22–25,000 times, Pfankuch and Kausche observed phage particles appearing as round particles of 60 nm in diameter. These observations were confirmed by Helmut Ruska and he proposed to name those morphological complex particles as "Herellen," which had club-like structures with heads of 90–110 nm and tails of 25–250 nm in length.[6]

UNDERSTANDING THE GENETIC MATERIAL (NUCLEIC ACIDS)

Ancillary structures supported the virus as being made up of genetic material (nucleic acids). A German biochemist, Ludwig Karl Martin Leonhard Albrecht Kossel (September 16, 1853–July 5, 1927) won the Nobel Prize for Physiology or Medicine in 1910 for isolating and determining the chemical composition of nucleic acids: adenine, cytosine, guanine, thymine, and uracil (nucleic acid building molecules).[7] But the platform kept these building molecules in a particular configuration.

Phoebus Levene, a Russian immigrant to America, gained his MD from St. Petersburg in 1891, enrolled in classes at Columbia University, New York, and subsequently joined the Rockefeller Institute for Medical Research in 1905.[8] He discovered that the platform was composed of sugar, which helped building molecules together in an organized manner. But the sugar in nucleic acid was not ribose as it lacked one oxygen atom—so he called it deoxyribose. Subsequent research would divide nucleic acid into deoxyribonucleic acid (DNA) and RNA.[8] RNA is single-stranded (ss) and has the sugar ribose, and DNA is double-stranded (ds) and the deoxyribose sugar. Primary storage of genetic information for all cells is deoxyribonucleic, which stays in the nucleus. RNA molecules travel from the nucleus to the cytoplasm to assist in the execution of the instructions from DNA. Synthesis of RNA from one strand of DNA is called transcription. The genetic information from DNA is carried by messenger RNA to the ribosomes to synthesize new functional proteins. Unlike living cells, RNA may only be the bearer of its genetic material in the viruses and are therefore called RNA viruses.[8]

STRUCTURE OF VIRUSES

All viruses contain nucleic acid (the genetic material), either DNA or RNA. Nucleic acid is enclosed by a protein envelope, a covering made from combination of fat and proteins that can also surround the whole virus. A virion is a virus particle outside the host cell but still capable of infecting other cells. In a virion, nucleic acid can produce specific proteins that can serve as infecting agent for the virus. Proteins strongly integrated with nucleic acid can form a virus-like particle known as "prions."

The basic viral structure is composed of the following three components:

Nucleic acid—Genetic information required for the production of proteins and other structural and function components is stored in nucleic acid. Only a few groups of viruses use DNA. Most viruses preserve all their genetic information with the ssRNA. There are two types of RNA-based viruses: a plus or positive-strand RNA can directly serve as a code in the formation of proteins and enzymes needed by the virus; a negative-stranded RNA needs to be processed by an enzyme to form messenger RNA, which then can be used for protein synthesis.

Capsid—Nucleic acid is enclosed in a shell of proteins called a capsid. Proteins are arranged in subunits known as capsomeres and then tightly integrated within nucleic acid giving it a shape of polygon-like sphere or a helix. Functionally, the capsid has three roles:

1. It serves as a shield for the genetic material against the action of digestive enzymes.
2. With the help of some special sites in its structure, it aids in attachment to the host cell membrane.
3. It contains enzymes, or proteins, enabling the virion to penetrate host cell membranes and transports nucleic acid inside the cells.

The capsid enclosing nucleic acid is referred to as nucleocapsid, which can be considered as an infectious and functional virus.

Envelope—In some viruses, the nucleocapsid is covered with another layer of proteins and fats called as glycoproteins. This envelope is composed of proteins sandwiched between two layers of lipids or fats. Sometimes, it can also contain parts of host cell membranes as virus leaves host cells during the budding process. Lipids or fats are generally taken from host cells, and the virus incorporates its own manufactured proteins in it to provide structural stability. Glycoprotein spikes can also be present on the envelope, which help to anchor the virus into host cell membranes.

VIRUS AND HOST CELL

The interaction of viral and host cell nucleic acid is mandatory in order to manufacture proteins needed for virus replication and other functions. Viral nucleic acids enter the nucleus of host cell, where the DNA is decoded into the messenger RNA. This process is called transcription. After transcription, messenger RNA makes its way out of nucleus to the cytoplasm where with the help of ribosomes it utilizes amino acids to construct proteins. A simple way to think of this process is to think of virus nucleic acid as a USB disc that attaches to the computer (nucleus) and the data are integrated with the program of the computer, sent to the printer through cable (RNA), and the printer (ribosomes) prints out the data (proteins).

EBOLA VIRUS STRUCTURE

Having discussed the viral structures and components in general, the following section will focus specifically on Ebola virus structure, mechanism of entry into human cells, replication and immune system evasion.

The Ebola virus is a member of Filoviridae family. Viruses in that family consist of single, linear, negative-sense ssRNA genomes. The Filoviridae family gets its name from the Latin work meaning *filum,* meaning thread-like, because viruses in this family have thread-like appearance under an electron microscope. *Cuevavirus, Ebolavirus,* and *Marburgvirus* are the three genera that are placed under this family.[9] Baltimore classification system was devised by David Baltimore, a biologist who won Nobel Prize. Current virus classification is based on the International Committee on Taxonomy of Viruses classification system and the Baltimore classification system. The Baltimore classification of viruses is based on the mechanism of messenger RNA (as discussed earlier). Messenger RNA can also has the ability to replicate itself. Viral genomes may be ssRNA/DNA or dsRNA/DNA. Viruses can or cannot use reverse transcriptase.[10]

Dr Peter Piot, codiscoverer of the Ebola virus, wrote the details of how the Ebola virus got its name in his memoir "No Time to Lose." In 1976, a team of scientific investigators was working in pursuit of Ebola virus disease outbreak. They decided to name the virus that had already taken a large number of lives after the village where the virus began: Yambuku. Dr Joel Breman, who was working for Center for Disease Control and Prevention (CDC) objected to this idea because naming the virus Yambuku would spread the terror in the village and the village could become a symbol of devastation. In Nigeria in 1969, Lassa virus was named after it emerged in a village called Lassa and it created many problems for the settlers. Another researcher, Karl Johnson, who was also leading the research team, suggested to name the virus after a river, to obscure the focus on a specific area. The team of investigators decided to look at the map and name the virus on the name of closest river. So on the map it appeared that Ebola River was the nearest one to the Yambuku village. In local language, Ebola means "Black River". Later, it was realized that the map they used was not accurate and the Ebola River was not the closest river.[11]

The actual structure of Ebola virus was described by two groups supervised by Dr K.M. Johnson and Dr Stephan Pattyn. Dr Johnson at CDC studied the tissue samples of patients with Ebola virus infection. During the Ebola virus disease outbreak in Southern Sudan and Northern Zaire, Dr S Pattyn sent blood and tissue specimens to the CDC in Atlanta, USA. There Dr Johnson and his team inoculated these specimens into African green monkey cells. Under electron microscope, large number of filamentous virus particles were visualized. They were 100 nm in diameter, and length varied from 300 nm to more than 1500 nm. These particles had 10-nm-long projections on their surface. Dr Johnson described these particles as indistinguishable from Marburg virus particles studied in 1967 (isolates from Germany) and 1975 (isolates from South Africa). They noticed two prominent characteristics in the 1976 Zaire isolated specimens: there was more branching of filamentous particles and the envelope continued beyond the ends of rigid internal structure. The diameter of Ebola virion is around 80 nm and the length is 800 nm.[12] In the center of the particle is the viral nucleocapsid which consists of the helical RNA genome wrapped around the viral proteins. Glycoproteins are a class of proteins that have carbohydrate groups attached to the polypeptide chain. Between the capsid and envelope are viral proteins 40 and 24.[13]

Ebola virus helical SS-RNA has a sequence of genes in a specific pattern. These genes are responsible for the formation of virus proteins. The gene releases mRNA within the host cell and mRNA integrates with the genes of host cells. This alteration in host cell genetic material results in the production of viral proteins within the host cell. A special glycoprotein is formed within the host cell and expressed at the cell surface. This envelope protein is responsible for further membrane fusion and viral attachment. It has also been shown as the crucial factor for Ebola virus pathogenicity. Pathogenicity can be defined as the disease-causing ability of the virus. Glycoproteins are actually processed by the enzyme known as protein convertase furin, which leads to the formation of two glycoprotein 1 and glycoprotein 2 subunits. Glycoprotein 1 allows for attachment to host cells, while glycoprotein 2 mediates fusion of viral and host membranes.[14] This protein on the viral envelope plays an important role to merge the two membranes.[15]

Two proteins are located under the viral envelope and help maintain structural integrity. One of them, viral protein 40, is involved in filovirus budding by inducing its release from cells in the absence of all other viral proteins.[16] The second matrix protein, viral protein 24, has been shown to suppress interferon (IFN) production. IFNs are chemicals produced by human cells in response to infection or foreign invaders. Viral protein 24, along with viral protein 35 and NP, are sufficient to form nucleocapsid structures.[17] Nucleocapsid is the virus genetic material enclosed in a membrane of proteins. It is essential for virus function that all the components and proteins that form nucleocapsid must be arranged in the right manner, which is where viral protein 24 plays its role. Viral protein 35, NP, and RNA polymerase L are some other proteins that, along with viral protein 24, function to form nucleocapsid. These NP proteins function in dual way during the viral replication cycle. They not only form structural components, but also replicate and transcribe the genome. As evidenced by all of these proposed mechanisms, viral protein 35 makes Ebola virus a lethal virus. A second matrix protein, viral protein 24, suppresses the natural defense mechanism of IFN production as well. Normally, signal transducer and activator of transcription complexes or STAT complexes are transported to the nucleus where they initiate signals for the production and release of IFN. Viral protein 24 disables the STAT transport system to the nucleus of cell, thus rendering human cells unable to produce IFN.[18]

VIRAL ENTRY INTO THE HOST CELL

The process of viral infection starts with the entrance of the virus into host cells. This happens when the host cells come into direct contact with virus, which inserts its genetic material into the host cells. The Ebola virus must enter the host cell to survive and replicate. How Ebola enters the host cells is not exactly known yet. Many enveloped viruses, similar to Ebola virus, rely upon the endocytosis to enter cells. The host cells engulf virus as if they are food particles. This process of engulfment can be referred as endocytosis. Cells that utilize surface receptors to transport resources from the environment into the cells usually have the potential to engulf viruses as well.

While the cell membrane was initially viewed as a randomly arranged lipid bilayer, now cholesterol-enriched regions known as lipid rafts have been identified. These lipid platforms also serve as a site for the intimate reactions (landing sites) of Ebola virus proteins required for entry and departure.[19] These lipid rafts (landing sites) may also impact viral pathogenicity and host immune response by acting as a catalyst. Lipid rafts contain molecules that can merge into the envelope of budding virions and used to ensure more efficient replication.[19] Ebola virus enters the cells through large, heterogeneous vesicles known as macropinosomes. The process starts with outward extensions of the cell membrane (outer layer of the cells) around the virus. Distal ends of the membrane fuse leading to the formation of a ring "macropinosome" around the virus.[20]

The process appears like a person greeting someone with arms open and subsequently hugging the visitor. Even though the process is not completely understood, it most likely involves the interaction of viral surfaces of glycoproteins with cell-surface receptors. Most viruses rely on the presence of specific proteins, clathrin (CME), and caveolin (CavME) for endocytosis. At the University of Texas Medical Branch, Galveston, Dr Mohammad F. Saeed and his team of researchers demonstrated that the Ebola virus can enter cells even when the production of these proteins was inhibited.[21] After making entry into the cell, the virus now exits in the ring or macropinosome and starts its process of replication and infection.[22] Low pH in the cytoplasm is the factor that signals the Ebola virus to come out of vesicle.[23] The virus then moves into the cell cytoplasm at a site where replication optimally proceeds.

EBOLA VIRUS TRANSCRIPTION

Transcription is the process of copying the genetic code of the virus into the mRNA of the host cells (interaction between genetic material of virus and host cell), which can then work with a set of enzymes to form viral proteins. This messenger RNA can also serve as a code to replicate the virus's genetic material. Polymerase complexes bind special sites in the genome to initiate the transcription process. The complex then slides along the RNA and sequentially transcribes the individual genes in a particular order.[24]

The process resembles fingers moving over a set of piano keys. Each key represents a single gene in this process of stepwise interaction. Polymerase complexes are enzymes or molecules that assist in the process of replication. Viral protein 30 is considered an important protein in the transcription process crucial to virus life cycle. Interestingly, there are certain viral proteins that inhibit its own replication and transcription. This interference could be significant in converting the viral genetic structure from a transcription or replication-competent form to one that is ready for virion assembly.[25] Viral budding is the phenomenon by which virus particles leave infected host cells. These particles use host cell membranes as their temporary home and leave to infect more cells and organs in the body.

THE BODY'S IMMUNE SYSTEM

White blood cells

White blood cells, also known as leukocytes or leucocytes, are the backbone of the immune system responsible for protecting against all harmful agents that infect the human body. In mid-eighteenth century, William Hewson was among the pioneers to identify white blood cells. He used the term "colorless cells" for white blood cells. He described that the white blood cells are produced in lymphatic system, where from they are transported to the blood circulation via a specialized duct system. He further proposed that, in blood vessels, white blood cells are converted into red blood cells. He used serum instead of water to dilute the blood and further investigate white blood cells.

During that time, stains were not yet developed so that he could study white blood cells in more detail. It was not until the nineteenth century, when Paul Ehrlich stained cells using dyes, which helped him to differentiate among various types of blood cells and red cell changes in diseases and also identified the various types of white blood cells. Elie Metchnikoff discovered neutrophils when he pricked starfish larvae with a rose thorn and noticed the accumulation of these cells at the site of injury.

Generally, the mode of transmission for Ebola virus is direct contact with skin or body fluids of infected human or animal. White blood cells respond by releasing large amounts of proinflammatory cytokines. If leukocytes are considered police officers, proinflammatory cytokines will be the bullets used to stop the perpetrator virus. Unfortunately, proinflammatory cytokines (almost like stray bullets hitting normal habitat) damage and increase the leakiness of the blood vessels, facilitating easier entry into the virus's secondary targets, endothelial cells. Ebola virus causes considerable damage to liver cells, which lose normal functioning adding to a plethora of problems.[26]

INTERFERON PRODUCTION

IFNs are proteins manufactured and released by our cells when they get infected with agents such as viruses, bacteria, parasites, or tumor cells. IFNs protect against viral infections by interfering with viral replication. If proinflammatory cytokines are bullets used by leukocytes, IFNs are the bullets used by the host cells that are falling victim to the virus. Normally, IFNs are released from cells infected with the virus. That, in turn, enhances defense mechanisms against viral invasion.

Alick Isaacs and Jean Lindenmann were the two scientists who described IFNs at the National Institute for Medical Research in London who designed an experiment to inhibit the proliferation of live influenza virus in chicken embryo. This experiment showed that a protein inhibited the growth of live influenza virus. Due to its ability to interfere viral replication, in 1957, they named that protein an "interferon." Many viruses have devised mechanisms against this antiviral function of IFNs. They not only stop IFN production, but also disable proteins that are activated by IFNs.[27]

EBOLA VIRUS MESSING WITH THE BODY'S IMMUNE RESPONSE

Ebola virus is unique because of its ability to interfere with or avoid the human immune system. During infection, leukocyte subtypes called monocytes are the cells that get attacked by the virus. Since these cells are very important in the acute phase of inflammatory response against the virus, their incapacitation allows the virus to evade early containment and destruction. Nitric oxide is released in an excessive amount from infected macrophages. Nitric oxide is a gaseous hormone that normally functions in cell communication. However, in high concentrations, nitric oxide causes programmed cell death (apoptosis) in adjacent natural killer cells, further weakening the initial response to the virus.[26] Large amounts of proinflammatory cytokines secreted from macrophages lead to massive damage in blood vessels as mentioned previously.

DISRUPTION OF BLOOD VESSELS AND VIRUS ENTRY INTO THE TISSUE

The wall of the capillaries is sealed by a layer of endothelial cells bound by extracellular matrix just like a layer of bricks held together by cement. Ebola virus disrupts the endothelial cell, making the wall leaky and allowing blood to seep outside. The expression of Ebola virus glycoproteins in cultured cells causes a disruption in cell adhesion, resulting in a loss of cell-to-cell contact. Ebola virus infections are associated with an alteration of surface (cell membranes) of the inner lining of cells that break open the existing binding between cell and extracellular matrix.[28,29]

THE FUTILE LEUKOCYTE CHASE

Leukocytes move out of blood vessels to the location of tissue damage in a movement known as extravasation. Pathogens activate macrophages in the infected tissue, which then release certain chemicals called cytokines. These cytokines send signals to the endothelial cells (layer of cells that form the inner lining) of blood vessels near the site of infection. The endothelial cells express adhesive molecules on their surface called selectins and become sticky. Chemokines signal white blood cells to migrate toward the area of virus infection where they stick to the blood vessel walls. Leukocytes pass through spaces between blood vessel cells and the process from attachment to transport across the wall of the blood vessels is called diapedesis. Diapedesis is followed by movement of leukocytes toward the areas of infection marked by high concentration of inflammatory proteins. The small glycoproteins released by virus-infected cells prevent leukocytes from entering the tissue (transmigration process) to attack the virus. Leukocytes appear to stick within the blood vessels, unable to reach the site of infection.[30]

Stuck leukocytes continue to release a cocktail of proinflammatory cytokines that destroy the surrounding blood vessel walls, but also activate the coagulation cascade. It is as if the police (leukocytes) has arrived through the roads alerted by the noise and commotion (chemokines and cytokines) inside the house but the doors of the house are stuck (endothelial cell barriers) due to glue (glycoprotein) injected by the perpetrators (virus) forcing the police to gather outside on the road watching helplessly as the perpetrators (virus) continue their destruction unchecked. The gathered police continue to fire bullets (proinflammatory cytokines) into the walls, which cause a lot of damage to the house but do not prevent destruction that continues within the house.

SUMMARY

Viral replication and defective immune responses make Ebola virus infections lethal. Delay in defensive responses against early Ebola virus infections is responsible for this catastrophe. Blood transports the virus to different tissue in the body. The virus makes its way through the blood vessels to many organs where it causes all the damage. For many reasons, blood vessels repair themselves and thus do not allow white blood cells to pass to the locations where they can counteract the virus.

REFERENCES

1. Lustig A, Levine AJ. One hundred years of virology. *J Virol* August 1992;**66**(8):4629–31. PubMed PMID: 1629947. Pubmed Central PMCID: PMC241285. Epub 1992/08/01. eng.
2. Kyle RA, Shampo MA. Theodor Svedberg and the ultracentrifuge. *Mayo Clin Proc* September 1997;**72**(9):830. PubMed PMID: 9294529. Epub 1997/09/19. eng.
3. Beams HW, Bronte Gatenby J, Bluliyil JA. Ultra-centrifuging the spermatocytes of helix aspersa. *J Cell Sci* 1936. [cited 12/16/2014]. Available from: http://jcs.biologists.org/content/s2-78/311/387.full.pdf+html.
4. Pennazio S, Roggero P. The discovery of the chemical nature of tobacco mosaic virus. *Riv Biol* May–August 2000;**93**(2):253–81. PubMed PMID: 11048483. Epub 2000/10/26. eng.
5. Fowden L, Pierpoint S. Norman Pirie (1907–97). *Nature* June 5, 1997;**387**(6633):560. PubMed PMID: 9177338. Epub 1997/06/05. eng.

6. Goldsmith CS, Miller SE. Modern uses of electron microscopy for detection of viruses. *Clin Microbiol Rev* October 2009;**22**(4):552–63. PubMed PMID: 19822888. Pubmed Central PMCID: PMC2772359. Epub 2009/10/14. eng.

7. www.nobelprize.org. The nobel prize in physiology or medicine 1910–2014 [cited 12/16/2014]. Available from: http://www.nobelprize.org/nobel_prizes/medicine/laureates/1910/.

8. Tipson RS. Phoebus Aaron Theodor Levene, 1869–1940. *Adv Carbohydr Chem* 1957;**12**:1–12. PubMed PMID: 13617111. Epub 1957/01/01. eng.

9. Kuhn JH, Becker S, Ebihara H, Geisbert TW, Johnson KM, Kawaoka Y, et al. Proposal for a revised taxonomy of the family Filoviridae: classification, names of taxa and viruses, and virus abbreviations. *Arch Virol* December 2010;**155**(12):2083–103. PubMed PMID: 21046175. Pubmed Central PMCID: PMC3074192. Epub 2010/11/04. eng.

10. Viruses ICoTo. *International Committee on Taxonomy of Viruses.* 2014. [cited 17/12/2014]. Available from: http://ictvonline.org/.

11. Loria K. *Scientists who discovered Ebola almost caused a disaster: 'It makes me wince just to think of it'.* 2014. [cited 12/16/2014]. Available from: http://www.businessinsider.com/peter-piot-tells-the-ebola-discovery-story-2014-8.

12. Johnson KM, Lange JV, Webb PA, Murphy FA. Isolation and partial characterisation of a new virus causing acute haemorrhagic fever in Zaire. *Lancet* March 12, 1977;**1**(8011):569–71. PubMed PMID: 65661. Epub 1977/03/12. eng.

13. Feldmann H, Klenk HD, Sanchez A. Molecular biology and evolution of filoviruses. *Arch Virol Suppl* 1993;**7**:81–100. PubMed PMID: 8219816. Epub 1993/01/01. eng.

14. Volchkov VE, Feldmann H, Volchkova VA, Klenk HD. Processing of the Ebola virus glycoprotein by the proprotein convertase furin. *Proc Natl Acad Sci USA* May 12, 1998;**95**(10):5762–7. PubMed PMID: 9576958. Pubmed Central PMCID: PMC20453. Epub 1998/05/20. eng.

15. Lee JE, Fusco ML, Hessell AJ, Oswald WB, Burton DR, Saphire EO. Structure of the Ebola virus glycoprotein bound to an antibody from a human survivor. *Nature* July 10, 2008;**454**(7201):177–82. PubMed PMID: 18615077. Pubmed Central PMCID: PMC2700032. Epub 2008/07/11. eng.

16. Jasenosky LD, Neumann G, Lukashevich I, Kawaoka Y. Ebola virus VP40-induced particle formation and association with the lipid bilayer. *J Virol* June 2001;**75**(11):5205–14. PubMed PMID: 11333902. Pubmed Central PMCID: PMC114926. Epub 2001/05/03. eng.

17. Huang Y, Xu L, Sun Y, Nabel GJ. The assembly of Ebola virus nucleocapsid requires virion-associated proteins 35 and 24 and posttranslational modification of nucleoprotein. *Mol Cell* August 2002;**10**(2):307–16. PubMed PMID: 12191476. Epub 2002/08/23. eng.

18. Reid SP, Leung LW, Hartman AL, Martinez O, Shaw ML, Carbonnelle C, et al. Ebola virus VP24 binds karyopherin alpha1 and blocks STAT1 nuclear accumulation. *J Virol* June 2006;**80**(11):5156–67. PubMed PMID: 16698996. Pubmed Central PMCID: PMC1472181. Epub 2006/05/16. eng.

19. Bavari S, Bosio CM, Wiegand E, Ruthel G, Will AB, Geisbert TW, et al. Lipid raft microdomains: a gateway for compartmentalized trafficking of Ebola and Marburg viruses. *J Exp Med* March 4, 2002;**195**(5):593–602. PubMed PMID: 11877482. Pubmed Central PMCID: PMC2193767. Epub 2002/03/06. eng.

20. Weed SA, Parsons JT. Cortactin: coupling membrane dynamics to cortical actin assembly. *Oncogene* October 1, 2001;**20**(44):6418–34. PubMed PMID: 11607842. Epub 2001/10/19. eng.

21. Saeed MF, Kolokoltsov AA, Albrecht T, Davey RA. Cellular entry of Ebola virus involves uptake by a macropinocytosis-like mechanism and subsequent trafficking through early and late endosomes. *PLoS Pathog* 2010;**6**(9):e1001110. PubMed PMID: 20862315. Pubmed Central PMCID: Pmc2940741. Epub 2010/09/24. eng.

22. Doherty GJ, McMahon HT. Mechanisms of endocytosis. *Annu Rev Biochem* 2009;**78**:857–902. PubMed PMID: 19317650. Epub 2009/03/26. eng.

23. Schornberg K, Matsuyama S, Kabsch K, Delos S, Bouton A, White J. Role of endosomal cathepsins in entry mediated by the Ebola virus glycoprotein. *J Virol* April 2006;**80**(8):4174–8. PubMed PMID: 16571833. Pubmed Central PMCID: PMC1440424. Epub 2006/03/31. eng.

24. Emerson SU. Reconstitution studies detect a single polymerase entry site on the vesicular stomatitis virus genome. *Cell* December 1982;**31**(3 Pt 2):635–42. PubMed PMID: 6297777. Epub 1982/12/01. eng.

25. Watanabe S, Noda T, Halfmann P, Jasenosky L, Kawaoka Y. Ebola virus (EBOV) VP24 inhibits transcription and replication of the EBOV genome. *J Infect Dis* November 15, 2007;**196**(Suppl. 2): S284–90. PubMed PMID: 17940962. Epub 2007/12/06. eng.

26. Geisbert TW, Hensley LE, Larsen T, Young HA, Reed DS, Geisbert JB, et al. Pathogenesis of Ebola hemorrhagic fever in cynomolgus macaques: evidence that dendritic cells are early and sustained targets of infection. *Am J Pathol* December 2003;**163**(6):2347–70. PubMed PMID: 14633608. Pubmed Central PMCID: PMC1892369. Epub 2003/11/25. eng.

27. Basler CF, Wang X, Muhlberger E, Volchkov V, Paragas J, Klenk HD, et al. The Ebola virus VP35 protein functions as a type I IFN antagonist. *Proc Natl Acad Sci USA* October 24, 2000;**97**(22):12289–94. PubMed PMID: 11027311. Pubmed Central PMCID: PMC17334. Epub 2000/10/12. eng.

28. Francica JR, Varela-Rohena A, Medvec A, Plesa G, Riley JL, Bates P. Steric shielding of surface epitopes and impaired immune recognition induced by the Ebola virus glycoprotein. *PLoS Pathog* 2010;**6**(9):e1001098. PubMed PMID: 20844579. Pubmed Central PMCID: PMC2936550. Epub 2010/09/17. eng.

29. Sullivan NJ, Peterson M, Yang ZY, Kong WP, Duckers H, Nabel E, et al. Ebola virus glycoprotein toxicity is mediated by a dynamin-dependent protein-trafficking pathway. *J Virol* January 2005;**79**(1):547–53. PubMed PMID: 15596847. Pubmed Central PMCID: PMC538691. Epub 2004/12/15. eng.

30. Wahl-Jensen VM, Afanasieva TA, Seebach J, Stroher U, Feldmann H, Schnittler HJ. Effects of Ebola virus glycoproteins on endothelial cell activation and barrier function. *J Virol* August 2005;**79**(16):10442–50. PubMed PMID: 16051836. Pubmed Central PMCID: PMC1182673. Epub 2005/07/30. eng.

Chapter 9

Clinical Manifestations and Laboratory Diagnosis of Ebola Virus Infection

Chapter Outline

A TYPICAL CASE OF EBOLA VIRUS DISEASE

A 33-year-old man was working in Liberia as a health-care worker since October 2013. He took his daily prophylaxis against malaria infection. By April 2014, he and his team had established an Ebola treatment unit in Monrovia, Liberia. In July 2014, he woke up feeling feverish and tired. He measured his temperature and found that he was slightly febrile with a temperature of 37.8 °C when measured through a thermometer placed in the mouth (the normal temperature is 37.4 °C). He informed his colleagues of his illness and stayed at his local residence in Liberia. Two rapid malaria tests were negative for identification of malaria infection. With the passage of time, his fever got worse, heightening to a body temperature of 38.6 °C. He also felt nauseated.

Polymerase chain reaction (PCR) tests for Lassa fever, yellow fever, and Ebola virus disease were all negative at the time. As his fever continued, he was treated with intravenous fluids and antibiotics. On day 4 of his illness, tests were repeated for malaria, Lassa fever, yellow fever, and Ebola virus. This time, the test demonstrated that he had Ebola virus disease. It is not uncommon for initial tests to be negative in very early Ebola virus infection because the amount of virus is too small to be detected in the blood.[1]

On day 6, he developed a petechial (spotty bluish) rash on his arms and chest. These rashes usually represent bleeding from small blood vessels under the skin. His fever spiked to an all-time high of 40 °C. He had severe abdominal pains and diarrhea. He had an episode of bloody stool and was subsequently transfused with a single unit of blood. Later on, his rash has turned into a maculopapular rash (a mix of flat and raised red bumps like lesions).[1]

On day 7, he was vomiting blood for which he was given another blood transfusion. He also received a unit of convalescent whole blood from an Ebola virus disease survivor. Despite the convalescent whole blood transfusion, his condition continued to worsen. For his fever and muscle pains, he received a gram of generic Tylenol every 6 h. He was given Tang and Gatorade to hydrate himself by mouth because of concerns regarding dehydration in the face of continuing vomiting and diarrhea. He appeared malnourished partly due to the severe anorexia.[1]

On day 9, a dose of intravenous Zmapp was administered, an experimental antibody-based treatment described in more detail in other chapters. Within 8 h of administration of Zmapp, he reported improvement in his energy level to the point that he was able to walk. The extent of his rash had decreased significantly as well. The professionals caring for him reported improvements in his fever and level of alertness.[1]

On day 10, he was transferred to Emory University's Communicable Diseases Unit in Atlanta, Georgia, for treatment. On day 11, he had slight fever with a temperature of 38.8 °C, heart beating fast with 120 beats per minute, and an oxygen saturation of 91%. Oxygen saturation is measured through a monitor taped to one of the fingers and measured the approximate concentration of oxygen in blood by measuring the wavelengths of light passed through arterial blood. The normal value is 95–99%. He was started on 2 L of supplemental oxygen administered via inhalation and resulted in improvement in saturation to 93%. Normally, a resonating sound is heard when lungs are percussed, but there was a dull sound coming from his lungs on percussion, indicating that the low oxygen reading could be a result of inadequate oxygenation of blood because the lungs were not working well. He was severely dehydrated, and low blood pressure and rapid heart rate were recorded every time he stood up. His physical examination showed no obvious bleeding but the persistence of the same petechial rashes. He described pain after a manual pressing of the abdomen during examination, specifically in the right upper quadrant. His liver was not enlarged.[1]

He had some abnormalities detected on laboratory analysis. His internationalized normal ratio (a test of coagulation parameters) was 1-3, which is elevated and means that blood coagulation was delayed. He was found to have low potassium in serum, commonly seen with diarrhea, which was normalized with oral supplementation. Intravenous and oral fluids were continuously administered to treat dehydration. He received oral protein drinks and a multivitamin pill to improve his nutritional status.[1]

On days 12 and 15 of his illness, he received additional doses of Zmapp without any side effects. On day 15, his fever disappeared. Between days 14 and 17, the watery diarrhea was resolving, and volume and frequency of his bowel movements were normalizing. On day 17, the diarrhea had completely resolved and he no longer required intravenous fluids.[1] Polymerase chain reaction (PCR) tests were performed on his blood and it came back consistently negative. After negative PCR tests, it was decided on day 29 to remove him from isolation unit. He was discharged home the following day.[1]

The case presented above displayed certain characteristics that are seen in patients suffering from Ebola virus disease. A typical case will occur in a patient with no significant past medical history. The patients may report having been in contact with a stranger suffering from Ebola virus disease or a sick loved one. Since exposure, the patient will usually begin displaying symptoms within 8–12 days, although there is a possibility that the virus infection may not show symptoms for up to 21 days. The disease begins abruptly, patients feeling vague symptoms, a general sense of being unwell, and warm or under-the-weather. Patient can transmit disease to the people around them when these features are present.[1]

Within a short period of time, typically 1–4 days after initial symptoms, the patient will begin to complain of high fever, chills and shakes, profound weakness, and muscle pain. Some may describe these new occurrences as intense shaking. Weakness is sometimes so severe that the patient cannot even walk to the restroom. The muscle pain can be so severe that the patient does not want to move. Early signs of infection can be evident in the blood as the white blood cell count becomes elevated. White blood cells are elevated because, after being exposed to the infectious agent, the body signals a boost to the defense system. The white count rise is a marker of bodily response to combat the virus.[1]

Immediately after these symptoms and around 4–8 days after the symptoms initially start, patients can develop diarrhea, vomiting, abdominal pain, red eyes, bleeding, and, rarely, a rash. Patients can lose 20 L of body fluid a day from the severity of the diarrhea and vomiting, causing severe dehydration. The nausea is so bad that the patient can barely take fluids or food by mouth, making replacement of lost fluids and dehydration a challenge. The laboratory measurement of blood urea nitrogen and creatinine provides an estimate of the severity of the dehydration. These substances can be thought of as normal wastes that the body makes. These wastes are delivered to the kidney by the fluid portion of blood. After these wastes reach the kidneys, the kidneys dump these out of the body into the urine.

When there is severe dehydration, as there is in patients with Ebola virus disease, these wastes are not delivered to the kidneys and build up in the blood, hence the reason for their elevation in patients with Ebola virus disease. Normal electrolytes such as sodium, potassium, and chloride tend to be decreased in patients with Ebola virus disease due to enormous loss of fluid when the patients are constantly vomiting and having diarrhea. Potassium is mostly lost with diarrhea, and chloride is mostly lost with vomiting.[1]

About one in five patients may progress to experience massive bleeding between 6 and 16 days, which can manifest as blood in vomit or blood in diarrhea, making dehydration even more severe. The main blood cell responsible for blood clotting, platelets, start decreasing in the body due to Ebola virus disease. Platelets normally serve to initiate a temporary plug at a site of bleeding. Without platelets, there is a higher chance of bleeding, which explains the bleeding symptoms in Ebola virus disease patients.

At this point in the virus, blood vessels are damaged and a severe drop in blood pressure can lead to multiorgan failure known as hypovolemic shock. This is the point at which the patient is nearing his or her death. Signs of liver failure may be evident in blood tests with accumulation of normally produced liver enzymes such as aspartate aminotransferase (AST) and alanine aminotransferase (ALT), which are normally made inside liver cells. When liver cells are diseased, the cells die and dump AST and ALT into the blood, increasing the concentration of these enzymes. The elevation in Ebola virus disease patients may be either due to the direct attack of virus of these cells or severe dehydration that does not allow for enough blood to be delivered to the liver cells, causing them to die.

EBOLA VIRUS INFECTION AND SPREAD WITHIN THE BODY

A small group of invaders found their way through the walls of the fort. Isolated and too small in numbers to cause any commotion, they moved silently to overwhelm and confiscate the resources of the unsuspecting defending army's mobile units. Although the invaders were not going to receive any reinforcements, they were multiplying in numbers within the comfort of the confiscated resources at a rate that was unprecedented and out of this world. As the invaders rapidly traveled through the country under the guise of their victims, the defending armies were unable to mount a targeted offense to contain and destroy the invaders. The communications between defending units remained sporadic and ineffective. When the defending armies responded with massive artillery barrage, but without clear targets, the damage inflicted upon their own comrades exceeded the damage suffered by the invaders. The towns were left vulnerable and were quickly overrun by the ever-increasing numbers of invaders; these turned into ghost towns devoid of life and hope. Such was the story of the Ebola virus invaders as they unleashed themselves in the face of a disintegrated and ineffective response by armies of healthy cells.

Ebola virus can enter the body through many pathways: it can be a tear in skin or mucous membranes or it can be transferred from a blood transfusion infected with virus. Once in the body, Ebola virus can infect a large spectrum of cells including circulating white blood cells (that form the inner lining of several organs and blood vessels), fibroblasts and epithelial cells (important components of skin and muscles), and hepatocytes and adrenal cortical cells (secreting cells within the liver and adrenal glands).[2]

Usually circulating white blood cells, such as macrophages and dendritic cells, are the first cells to be infected irrespective of the point of entry into the human body. These cells become the vehicle of viral transport into various organs but, unfortunately, are not immune to the destructive effect of the virus. Ebola virus replicates rapidly within these cells, leading to their deaths with subsequent releases of large numbers of new viral particles into the extracellular fluid upon cell destruction.[3,4] Regional lymph nodes, which usually function to trap any infectious agent, capture the virus but things get worse when virus continues to replicate in lymph nodes, eventually leading to the destruction of this important defense line.

After that virus continues its rampage by taking down some other warriors of the body, like dendritic cells and macrophages, ultimately spreading to important organs such as liver, spleen, and kidney. Important protective arms of human body such as interferon and lymphocytes are suppressed by Ebola virus, which gives the virus the liberty to replicate and attack any site with any level of severity.[4] Studies have shown that the infected dendritic cells fail to mature and cannot produce antibodies. Lack of support signals from dendritic cells cause lymphocytes to undergo self-destruction that makes the body defense system even worse. With the progression of the disease, adrenal cortical cells, hepatocytes, fibroblasts, and many other cell types also become infected, resulting in extensive cell death and tissue necrosis.

The body unfortunately mounts a chaotic defense using expression and secretion of certain chemicals produced from macrophages infected with Ebola virus that are toxic to living cells. Peripheral blood samples have been taken from macaques infected with Ebola virus and these chemicals, along with some other compounds, have been found in an increased amount. Necrotic cells also signal the release of such chemicals. This response kills more of the body's cells than the virus, resulting in extensive tissue damage and inducing a systemic inflammatory syndrome.[3,4] High concentrations of such compounds in the blood lead to dilation of blood vessels, pooling of blood into various blood vessels, and exudation of fluid into the tissue, causing life-threatening loss of blood pressure and subsequent multiorgan failure as adequate perfusion to such organs is not possible. The condition is also referred to as septic shock to denote the etiological basis of a profound drop in blood pressure. Vomiting is another bodily defense mechanism and leads to loss of body fluids. Due to vomiting, patients are unable to take oral fluids and the psychological stress causes loss of appetite. In addition to the above-mentioned inflammatory substances, these are some other important factors that can contribute to the critical decrease in blood pressure.

The Ebola virus infection is noted for excessive bleeding as blood is unable to clot and thus the term "hemorrhagic fever" is used. Coagulation requires a combination of circulating proteins and adequately functioning blood cells called platelets. Platelets are specialized cells that function to prevent bleeding in the body. Ebola virus infection macrophages also produce some chemicals medically known as cytokines. Cytokines also induce production of tissue factors. Cytokines and tissue factors cause inappropriate and false activation of platelets and other proteins that are involved in blood clotting. This false activation leads to blood clotting that ultimately consumes a lot of platelets and other proteins. Due to depletion of platelets and proteins involved in normal clotting process, the body becomes prone to excessive bleeding and hemorrhages. As the disease progresses, liver injury may also cause a decrease in plasma levels of certain coagulation factors. This whole sequence is called disseminated intravascular coagulation.

INCUBATION PERIOD

In medicine, the time from the moment of exposure until signs and symptoms of the disease appear is called the incubation period. Onset of symptoms typically occurs 8–12 days after exposure to Ebola virus.[2,5] Patients who do not have symptoms of infection but are in their incubation period are not believed to transmit infection. However, there is clear evidence that virus can be communicated to other people from a symptomatic patient; therefore, precautions must be taken against exposure if a person is symptomatic with Ebola virus infection.[6] The clinical manifestation of Ebola virus disease starts with nonspecific influenza-like symptoms and progresses into a life-threatening picture with the involvement of multiple bodily functions and organs. The most common signs and symptoms reported from West Africa during the 2014 outbreak included diarrhea, vomiting, fatigue, fever, and anorexia or lack of appetite.[2]

CLINICAL MANIFESTATIONS

Ebola and Marburg virus infections have similar clinical manifestations in humans. Marburg virus is considered to be less lethal in terms of case fatality and severity of symptoms. Bleeding is not a common feature of Ebola virus disease, with most patients' only bleeding during the terminal phase of the infection.

The term "Ebola hemorrhagic fever" was replaced with the term "Ebola virus disease" in modern-day scientific literature. There have been many instances where Ebola virus was confused with another disease that had similar signs and symptoms. Initially, in 1976 when the first Ebola virus disease outbreak occurred, it was thought to be Marburg virus due to the fact that an outbreak had occurred the prior year. In other instances, Ebola virus infection was thought to be either Lassa fever or Yellow fever due to the fact that these diseases have similar symptoms. Back in 1995, during the Kikwit outbreak, many patients were believed to be infected with *Shigella*, which is a bacterium that can also cause bloody diarrhea.[7] Clinical characteristics of Ebola virus disease are discussed below. See Table 9.1 for a comparison of clinical symptoms in Ebola virus disease patients from Donka National Hospital in 2014, Kenema Government Hospital in 2014, and Kikwit General Hospital in 1995.[8]

TABLE 9.1 Demographic and Clinical Characteristics of Hospitalized Patients With Ebola Virus Disease at Donka National Hospital, Conakry, Guinea (March 25–August 5, 2014), Kenema Government Hospital, Sierra Leone (May 25–June 18, 2014), and Kikwit General Hospital, Democratic Republic of Congo (January 6–July 16, 1995)

Characteristics	Donka National Hospital 2014	Kenema Government Hospital 2014[9]	Kikwit General Hospital 1995[10]
Mean age (years)	34.0	NR	34.7
Men	63	42	47
Fever	66	89	NR
Headache	50	80	73
Weakness	79	66	78
Diarrhea	43	51	74
Abdominal pain	26	40	56
Vomiting	50	35	70
Conjunctivitis	9	31	34
Cough	4	20	NR
Rash	1	3	NR
Myalgia and/or arthralgia	20	NR	51
Anorexia	49	NR	73
Dyspnea	0	NR	25
Hiccup	9	NR	14
Any bleeding	19	NR	40–50[11]
Melena	6	NR	14
Hematemesis	3	NR	13

Abbreviations used: NR, not recorded; all values are represented as percentages.

Nonspecific flulike symptoms—Infection initially comes with symptoms consistent with common cold. Patients can have weakness, body aches, runny nose, fever with chills, and headaches. Lower back pain and aches in muscles surrounding the abdomen and back are also commonly seen.[12] Like typhoid fever, high-grade fever and decreased heart rate may also be noticed. The throat is inflamed and swollen, and patients complain of feeling a painful lump in the throat. At this stage, the infection cannot be differentiated from any other viral infection that involves the respiratory system including influenza.

Rash—By days 5–7 of the illness, a diffuse red rash may appear on the skin. The rash can cover massive body areas, starting from face and extending down to the trunk with some ulceration of skin as well.[7,12–14]

Gastrointestinal—Watery diarrhea, nausea, vomiting, and abdominal pain usually develop after many days of initial symptoms. The gastrointestinal symptoms and rash provide the first hints that the infection is more extensive than what would be expected in a respiratory viral illness.[14]

Hemorrhagic manifestations—Bleeding is not an early feature of the disease, not always present, and, therefore, may not be helpful in the diagnosis of the infection. In the later stages of the infection, excessive bleeding may be evident as ecchymosis/bruising, petechial, oozing from venipuncture sites, and mucosal hemorrhages from nose and gastrointestinal systems. Frank hemorrhages such as bleeding from the nose, rectum, and vagina are seen most commonly in the terminal phase of illness. In the current outbreak, not less than 20% of patients have suffered bleeding, with gastrointestinal tract bleeding being the most common.[2,14]

Myalgia, arthralgia, and fatigue—All viral diseases produce myalgia and fatigue of varying severity. The severe myalgia and fatigue may be secondary to dehydration and uremia seen in acute kidney failure. Poor nutritional status is another contributor to such manifestations. Antigen–antibody complexes are formed during the recovery phase, which may lead to acute arthralgia and other symptoms similar to those seen in autoimmune arthritis.[3]

Psychological symptoms—Ebola virus disease, like any other untreatable condition, has devastating psychological stress on patients. With its high mortality rate and no proven treatment, patients are more likely to suffer from depressive symptoms. Lack of family support due to the risk of exposure increases the load of mental health problems. Major depression, social withdrawal, loss of appetite, suicidal ideations, and attempts to commit suicide have been reported in patients suffering from Ebola virus disease.[15]

Miscellaneous findings—Ebola virus disease can present with findings such as difficulty in breathing, pain in the chest, headache, hiccups, confusion, seizures, and swelling in the brain. Redness of eyes and redness of the soft palate are common findings.[13] Several of these manifestations are a result of multiorgan failure rather than the direct effect of virus dissemination into muscles or the central nervous system. Spontaneous miscarriages may occur in pregnant women.

Patients demonstrate signs of improvement as early as first 6 days after the onset of symptoms in nonfatal infection.[2] Severe symptoms early on in the course of disease with progression to septic shock reflect a bad outcome in terms of survival. In severe Ebola virus disease, death ensues between days 6 and 16 predominantly due to shock.

LABORATORY DIAGNOSIS OF EBOLA VIRUS INFECTION

Because of the clinical presentation of Ebola virus disease being similar to several other viral and parasitic infections, diagnosis of Ebola virus infection depends on identifying sources of exposure and confirming virus antigens in serum. Fever in African countries is very frequently due to endemic malaria, typhoid, or Lassa fever and therefore, an unreliable symptom of Ebola virus infection. United States Centers for Disease Control and Prevention (CDC) and the World Health Organization (WHO) recommended some guidelines after the outbreak of Ebola virus disease in west part of Africa. Guidelines were provided to screen and manage persons who have been recently exposed to Ebola virus.[2,16–20] The approach depends on the severity of risk exposure and presence of sign and symptoms consistent with Ebola virus disease.

Because significant amounts of a sample of deoxyribonucleic acid are necessary for molecular and genetic analysis, studies of isolated pieces of deoxyribonucleic acid are nearly impossible without PCR amplification. The Nobel Prize was awarded to Kary B. Mullis, the creator of PCR. PCRs are very significant in a number of laboratory and clinical techniques, including detection of bacteria or viruses (particularly AIDS), DNA fingerprinting, and diagnosis of genetic disorders.

To amplify a segment of DNA using a PCR, DNA is first denatured by heating to separate into two pieces of single-stranded DNA. Then, an enzyme "Taq polymerase" uses the original strand as a template to synthesize two new strands resulting in duplication of original DNA. Two new strands can be created from each of these strands and so on. This cycle of denaturing and synthesizing new DNA is repeated to create billions of exact copies of the original DNA segment. The entire cycling process of PCR is automated, which can be completed in just a few hours. A machine called a thermocycler is designed to alter the temperature of the reaction in alternating manner to allow DNA denaturing and synthesis.

Reverse transcription PCRs (RT-PCRs) take only 3–10 days to detect Ebola virus after the onset of symptoms. The PCR is performed on the serum of suspected individuals. Reverse transcriptase can reverse the isolates and transcribe Ebola virus ribonucleic acid into deoxyribonucleic acid, which is then further amplified by the PCR, generating thousands to millions of copies of a particular deoxyribonucleic acid sequence.

Patients with symptoms of less than 3 days duration may need repeat testing.[21] Within the first 3 days, Ebola virus ribonucleic acid concentration may not be significantly high in blood circulation and obtaining a sample for viral detection may produce false-negative results. Enzyme-linked immunosorbent assay can detect viral antigens and specific ribonucleic acid sequences by RT-PCR.

Acute infections can be detected by RT-PCR. After the onset of symptoms, viral ribonucleic acid is generally detectable within 3–10 days by RT-PCR.[21] Sensitivity of RT-PCR can be affected by genomic variability and changes in ribonucleic acid sequence of Ebola virus.[22] Antigen detection may be used for an immediate diagnosis as a confirmatory test.[23]

The likelihood of exposure to Ebola virus and symptoms and/or basic laboratory tests suggestive of Ebola infection are the factors that determine which individual should be tested for Ebola virus infection.[2,16,20] Testing for IgG or IgM antibodies to Ebola virus can be used to identify to evaluate for past infection and/or immune response over time. PCR testing is also useful to determine whether a patient has recovered from Ebola infection. According to WHO, individuals without sign and symptoms of Ebola virus disease and two negative PCR tests on whole blood, at least 48 h apart, can be discharged as "infection free".[24]

ANCILLARY LABORATORY FINDINGS

Laboratory testing and phlebotomy should be limited to tests that are required for care.[25] Routine blood tests, such as complete blood counts, serum electrolytes, and urea and glucose, allow for quantification of the severity of dehydration and kidney and liver failure. Reduced levels of platelets and white blood cells may provide insights into the severity of blood coagulation impairment and immune suppression. Such tests may allow for early detection and monitoring of such manifestations and determine the response to therapeutic interventions.

Leukopenia—Decrease in number of white blood cells. It is followed by an increase in neutrophil cell count, with an increased percentage of immature cells. In 1967, Marburg outbreak caused many patients to have prominent decrease in the number of white blood cells with counts as low as 1000/μL at the time of clinical presentation.[2]

Thrombocytopenia—Platelet counts reach a nadir around days 6–8 of illness and are usually stay in the range of 50,000–100,000/μL in contrast to normal counts, which are usually in range of 150,000–350,000/μL.[2,12]

Elevated transaminases—Ebola virus invades hepatocytes (liver cells) and leads to necrosis of these cells. Elevated serum AST rises more than ALT levels. These enzymes are spilled in blood circulation when liver cells die due to viral invasion.[9,24]

Coagulation abnormalities—Fibrin degradation products are elevated, prothrombin and partial thromboplastin times are prolonged, consistent with disseminated intravascular coagulation. These changes are most prominent in severe cases of Ebola virus disease.

Renal abnormalities—Proteins in blood circulation do not filter through kidneys in urine. Due to damage to kidneys, plasma proteins start to seep through kidneys and excrete through urine of the patient. Renal insufficiency occurs with the progression of the disease.

FACTORS PREDICTING SURVIVAL OR DEATH AFTER EBOLA VIRUS INFECTION

Prognosis can be assessed on the basis of clinical and laboratory findings. Certain clinical manifestations such as increased respiratory rate, loss of urine formation, delirium, coma, and shock secondary to impaired oxygen delivery, intravascular volume depletion, and metabolic disturbances are associated with high fatality rate of the disease.[26] The clinical course of the disease also provides insights into the chances of recovery as clinical improvement is noted by second week in those who are likely to survive. In such patients, Ebola virus RNA concentrations drop and antibodies to Ebola virus appear during the second week.[27,28]

VIRAL PERSISTENCE

The virus can persist for some time in certain bodily fluids, including semen and breast milk. Persistence of virus in breast milk has also been noticed even after peripheral blood is tested negative for the presence of Ebola virus RNA.[29] Two children who were breast-fed by Ebola-infected mothers contracted and died of the disease even though the mothers were declared infection-free. Mothers of these babies recovered completely from Ebola infection, but during the course of infection they breastfed the babies.

Studies showed that reverse transcriptase PCR could easily detect viral RNA sequences in the semen of male patients for the period of up to 3 months, and infectious virus was recovered from one patient, 82 days after symptom onset.[30] During the 1967 Marburg outbreak, one case caused the transmission of virus through sexual contact because of viral persistence in semen.[13,31]

CONVALESCENCE OR RECOVERY PERIOD

It signifies the later stage of an infectious disease when the patient recovers, but may still be a source of infection even if patient started feeling better. Ebola virus disease has a prolonged convalescence period, which is characterized by extensive sloughing of skin and hair loss most likely due to necrosis of sweat glands.[28]

CDC GUIDELINES

Risk assessment of Ebola virus disease—Ebola virus disease risk should be evaluated if clinical finding suggestive of Ebola infection is present and there is a possibility of Ebola virus exposure within prior 21 days.[2,16,20]

Clinical findings—Severe headache, weakness, muscle pain, vomiting, diarrhea, abdominal pain, or unexplained hemorrhage are the symptoms that raise suspicion for Ebola virus disease.

Exposure—The level of exposure varies from high risk to low risk to no known exposure.

Individuals who are at risk for Ebola virus disease include:

- Contact with blood, body fluids, or any human remains of an individual known or suspected to have Ebola virus disease
- Residence in or the recent travel to an area with Ebola virus transmission
- Direct handling of rodents or primates from endemic areas.

As per the recommendations by the CDC, any of the following can be considered close contact:

- Being within the proximity of 3 feet or 1 m to an Ebola virus disease patient
- Being within the room of an Ebola virus disease patient for a prolonged duration without wearing protective equipment
- Contact with the patient while not wearing protection equipment
- Moving through a hospital, walking by a person, or providing care to a patient with Ebola virus disease.

Strict precautionary measures should be taken during the evaluation of patients suspected of Ebola virus infection. Suspected patients should be isolated in a room with a private bathroom, doors should be closed all the time, and health-care teams should wear protective gear to avoid any direct contact or transmission through droplets or body fluids. Gloves, face masks, gowns, and eye protection can be helpful in the prevention of virus transmission. The appropriate staff, the hospital infection control program, and state health departments should also be notified.

FURTHER EVALUATION

High-Risk Exposure—The CDC recommends testing for Ebola virus infection for all individuals with onset of fever within 21 days of having a high-risk exposure.[2,20] High-risk exposure persons without fever can have the same recommendation if there are other compatible clinical sign and symptoms present and/or laboratory findings are abnormal (i.e., raised transaminases and/or low platelet count, $<150,000$ cells/μL).

Any of the following includes high-risk exposure[24,25]:

- Mucous membrane or even percutaneous exposure to blood or certain body fluids of a person with Ebola virus disease
- Direct physical contact with skin, blood, or certain body fluids of an Ebola virus disease patient
- Processing blood or certain body fluids from a patient with diagnosed Ebola virus disease without appropriate personal protective equipment or standard biosafety precautions
- Having a direct contact with a dead body (including during funeral rites) without appropriate protective precautions in a country where an outbreak of Ebola virus disease is occurring.

Low-risk exposure—For persons with low-risk exposure who develop fever or other clinical findings consistent with Ebola virus disease, the CDC recommends some additional medical evaluation to check for Ebola virus disease.[16,19] The decision to test the individual is based upon the severity of illness, laboratory findings (e.g., platelet counts), and the likelihood of an alternative diagnosis.

A low-risk exposure includes[16,19]:

● Brief contact with an Ebola-infected patient without exposure to blood or other certain body fluids
● Another close contact with the Ebola virus disease patient in community settings or health-care facility.

Asymptomatic (or symptoms not meeting clinical criteria)—Symptoms and signs of Ebola virus disease should be monitored for certain asymptomatic individuals, as well as for those who do not meet the above clinical criteria. This includes individuals who have:

● Traveled in a country where an outbreak of Ebola virus disease has occurred within the past 21 days
● A high- or low-risk exposure.

Such individuals have to be monitored for 21 days after the last known exposure to a patient (or after leaving the country if no exposure is known) and should immediately report the onset of fever or other symptoms indicative of Ebola virus disease. Whether travel restrictions and/or monitoring by a public health authority are warranted depends upon the risk level of exposure, as described in the CDC recommendations for the monitoring and movement of individuals with Ebola virus exposure.[2]

WHO and CDC have almost similar recommendations. The WHO considers a person to be at high risk for infection if they have had sexual intercourse with a person recovering from Ebola virus disease or with a sick person.[17,19,32] PCR testing is also useful to determine whether a patient has recovered from Ebola virus infection. According to the WHO, individuals without signs and symptoms of Ebola virus disease can be discharged if they have two negative PCR tests on whole blood at least 48 h apart.[24]

BIOSAFETY AND BIOSECURITY CONSIDERATIONS WITH EBOLA VIRUS DISEASE

Direct interaction through skin abrasions or contact through mucous membranes of the nose and eyes can lead to spread of Ebola virus infection in health-care settings. Another source of spread is through exposure to blood or body fluids of Ebola virus disease-affected persons and through contaminated needles or syringes. According to the CDC guidelines, personal protective equipment can reduce the risk of self-contamination to health-care workers at risk of exposure. The description and levels of personal protective equipment are presented in

Tables 9.2 and 9.3. For effective implementation of these protective guidelines, the role of health-care facilities is essential. To safeguard health-care workers while taking care of Ebola virus disease patients, rigorous safety methods needed to be followed while putting on and taking off personal protective equipment.

TABLE 9.2 Types of Personal Protective Equipment in Health-Care Settings[33,34]

Types of Personal Protective Equipment	Designated Use
Gloves	Protects health workers' hands from exposure to infectious agents
Respirators and masks	Protects the respiratory tract from airborne infectious agents
Goggles	Protects the eyes of the health-care workers to reduce risk of potential splashing or spraying
Face shields	Protects face, mouth, nose, and eyes
Aprons and gowns	Protects the contamination of skin or clothing of health-care workers

TABLE 9.3 Levels of Personal Protective Equipment in Health-Care Settings[35]

Levels of Personal Protective Equipment	Indications	Examples
Level A	Required for the greatest level of protection to skin, respiratory tract, and eye	Positive pressure, full-face piece self-contained breathing apparatus, disposable protective suit, gloves, and boots
Level B	When highest level of respiratory protection is required	Positive pressure, full-face piece self-contained breathing apparatus, inner and outer chemical-resistant gloves, and face shields
Level C	Required when the concentration and type of airborne substances is known	Full-face air-purifying respirators, hard hat, escape mask, and inner and outer chemical-resistant gloves
Level D	Used when minimum protection is required	Gloves, safety glasses, chemical-resistant, steel-toe boots, or shoes

PRINCIPLES OF PERSONAL PROTECTIVE EQUIPMENT

According to the CDC guidelines, health-care workers must follow basic principles to ensure efficient use of personal protective equipment, most essential of which is the exposure of skin to virus, which should be avoided at all times. While entering the patient care area, it is essential to put on the personal protective equipment in a correct manner and order. It cannot be readjusted after entering the patient care area because any further manipulation with the protective equipment can result in breach of barriers against virus. A trained observer should monitor this activity to ensure proper compliance.

After handling the body fluids of the infected person, the gloved hands should be disinfected using an alcohol-based hand rub. During patient care activities, if the health-care worker experiences a partial or total breach of their personal protective equipment (through glove separation from sleeves leaving exposed skin, a tear in an outer glove, or a needle stick), then the health-care worker should be moved instantly to the designated area to evaluate the degree of exposure. If indicated during evaluation, the facility's exposure plan should be implemented emergently. In comparison to putting on protective gear, the removal of used personal protective equipment is a high-risk process that requires an organized approach, a trained observer, and a selected area for removal to warrant protection. This procedure should be slow and vigilant to reduce the probability of self-contamination.

During direct patient care and the personal protective equipment removal process, an additional layer of protection is provided by double gloving method. However, any further addition of layers will cause restraint in movements and lead to an excessively hot environment within the equipment, thus making patient care a challenging assignment. It is strongly suggested by the CDC to train and practice, in case any health-care facility decided to alter and enhance protective procedures with extra personal protective equipment. During the disposal process of used personal protective equipment, human error or omission can be expected. To prevent such occurrences, every step is read aloud to the health-care worker from the procedure worksheet, and each completed step is visually established and documented by a trained observer. The exclusive responsibility of a trained observer during entire put-on and takeoff process is to ensure the safety of the health-care worker and to make sure the facility is not compromised. The trained observer should have all the right knowledge about personal protective equipment recommendations documented in facility's protocol, including proper donning and doffing procedures, disposal of used personal protective equipment in order to provide right guidance to related health-care worker. In the event of an unintentional breach in procedure, the trained observer should be aware of exposure management plans.

Facilities should have a design that permits for a clear demarcation between clean and possibly contaminated areas. There should also be clearly visible signs to separate these distinct areas and to ensure a one-way movement of care from clean areas (area where personal protective equipment's donned and unused

TABLE 9.4 Categories of Pathogens According to Risk Group[40]

Risk Group Pathogens	Definition	Examples
Risk group 1	Microorganism causing no or low individual and community risk	*Escherichia coli, Bacillus subtilis,* Yeast
Risk group 2	Pathogen causing moderate individual risk, low community risk	Hepatitis B and C virus, human immunodeficiency virus (HIV), *Salmonella typhi*
Risk group 3	Pathogen causing high individual risk, low community risk	*Mycobacterium tuberculosis,* severe acute respiratory syndrome (SARS)
Risk group 4	Pathogen causing high individual and community risk	Ebola virus

equipment is stored) to the patient room, and then to the protective equipment removal area where equipment is removed and discarded.[36]

HANDLING OF BIOLOGICAL SPECIMENS SUSPECTED OF CONTAINING EBOLA VIRUS

While handling specimens from the patients, it is necessary for the laboratory staff to strictly follow safety protocol in order to prevent the transmission of bloodborne pathogens.[37] The only possible method to ascertain the etiology of a suspected case will be through a laboratory test, since the early signs of the Ebola virus disease can be vague.[38] Unquestionably, Ebola virus is a highly pathogenic agent; therefore, all laboratories should assess their resources and methods along with analyzing the training procedures and skills of their staff so as to conduct a comprehensive biohazard risk assessment. Implementation of correct procedures is necessary to lessen the possibility of the risk.[37]

Each of the laboratories should design a suitable workflow for processing suspected "Ebola virus-infected" samples to protect health-care providers and laboratory personnel. Specimens should be handled using an "enhanced chain of custody" algorithm. For example, having a designated staff for the shipping and supervision of the specimens within the laboratory is required to make sure that preanalytical, analytical, and postanalytical techniques are correctly tracked to amplify safety to personnel.[39] Ebola viruses are categorized as risk group-4 pathogens (see Table 9.4 for levels of pathogens), demanding Laboratory Biosafety Level (BSL)-4-equivalent control for isolation of virus.

Only if the virus has been inactivated in a BSL-3 facility (see Table 9.5 for definitions), an on-site molecular analysis can be done. BSL-4 or BSL-3

TABLE 9.5 Grades of Biosafety Level[41]

Biosafety Levels (BSL)	Criteria	Examples
BSL-1	Agents that pose minimal threat to laboratory workers; does not cause disease in healthy adults	Baculovirus
BSL-2	Agents causing human diseases, thus posing moderate hazard	Equine encephalitis viruses, human immunodeficiency virus (HIV)
BSL-3	Agents that are strictly controlled, can cause lethal disease via aerosol transmission	Yellow fever virus, St. Louis encephalitis virus, West Nile virus
BSL-4	Agents that are extremely dangerous and can pose a high risk of life-threatening disease	Ebola virus, Lassa virus

facility-lacking countries should assure the transportation of the specimens to an established WHO Collaborating Center.[22,38]

GUIDELINES FOR THE HANDLING OF LABORATORY SPECIMENS FROM SUSPECTED OR LABORATORY CONFIRMED EBOLA VIRUS DISEASE PATIENTS

- Personal protective equipment such as double gloves, impermeable laboratory gown, fluid-resistant gown worn over the laboratory coat or a combination of a sanctioned particulate respirator (e.g., N95 or N100), and eye protection (e.g., goggles/face shields/shroud), or a powdered air-purifying respirator use is strongly suggested, to be put on by the laboratory personnel while handling the samples from suspected Ebola virus disease patients.
- It is necessary to make sure adequate supplies of personal protective equipment and United Nations specimen triple packaging systems is currently available.[42]
- The samples collected from the suspected Ebola virus disease patients should not be handled in an open area.
- Procedures like pipetting, aspiration, slide preparation having the possibility to create infectious aerosols should be carried out in a biological safety cabinet in a minimum Containment Level-CL2 laboratory.
- The risk of aerosolization is higher with procedures like centrifugation of infectious material and should be carried out using sealed safety cups or sealed rotors that are only opened and unloaded in the biological safety cabinet.

- Use of glass or sharp instruments should be limited wherever necessary and laboratory personnel should be adequately trained in regular practices, including biosafety.
- Before collecting patient samples, label the tubes.
- Requisitions tested need to be stamped as "Ebola virus disease suspect," and labeling should be done on the exterior of the container.
- Make sure that all the specimens are safely stored and the accessibility should be limited only to authorized staff members.
- Trained staffs skilled in collecting blood should perform phlebotomy.
- During specimen processing, try to avoid unnecessary activities and minimize staff in that area.
- Prior to packaging for shipping to further testing, the surface of the specimen containers should be decontaminated by an effective disinfectant.
- To avoid contamination of the skin and hair, personal protective equipment should be removed in a proper manner. Any contact with the soiled and contaminated items (e.g., gloves, gowns, respirators) should be avoided.
- Proper sterilization is necessary for contaminated clothing and personal protective equipment before discarding, reuse, or removal from the laboratory.
- As soon as the personal protective equipment is removed, wash hands thoroughly.
- The spongy paper towels should be covered with disinfectant and left to soak for 15 min before cleaning up the accidental spills of potentially contaminated material.
- Specimens should be placed in a durable, leak-proof secondary container for transport within a facility. The external surface of a container should be disinfected with 3% sodium hypochlorite.
- Even for shipping within a facility, samples should be placed in a durable, leak-proof secondary container.
- Never use any pneumatic tubes for shipping the samples of suspected Ebola virus disease so as to avoid the risk of breakage or leaks.[43]
- All specimens should be stamped with a distinct patient ID. It should also be supplemented with a documentation sheet including at least patient's unique ID, date/time/place of sampling, type of specimen, test requirements, clinical data, along with travel history, and exposure to a suspected or confirmed Ebola virus disease case.[44]
- With regards to laboratory waste management during handling the specimens infected or contaminated with Ebola virus, the solid waste generated during laboratory testing should be steam sterilized (autoclaving), as the waste treatment process will inactivate the virus.[45]
- For handling Category A infectious substance like Ebola virus specimen, laboratories should have a Biological Safety Cabinet Certification.[45]
- If the specimen needs to be tested at the hospital laboratory which is the common laboratory for other hospital and patient-related activities, then the specimen should be double bagged, placed in a biohazard transportation container, and hand-carried to the laboratory[37] (Figure 9.1).

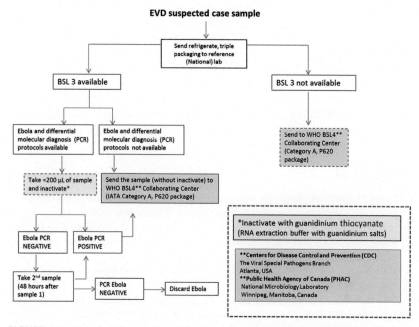

FIGURE 9.1 Flowchart for handling of samples derived from suspected Ebola virus disease patients. *World Health Organization (WHO). Algorithm for handling of samples from suspected Ebola virus disease. 2014.*

BIOSAFETY RECOMMENDATIONS FOR LABORATORIES CONDUCTING DIAGNOSTIC TESTING FOR EBOLA VIRUS DISEASE WITH APPROPRIATE BSL-3/BSL-4 FACILITIES

- A virus should only be isolated in a maximum containment BSL-4 laboratory. To prevent accidental or deliberate release of virus isolates, make sure the specimens are properly handled and safely stored.
- According to the detection protocol, the specimens should be inactivated only under BSL-3 conditions.
- Testing procedures like RT-PCR and enzyme-linked immunosorbent assay (ELISA) for noninactivated samples can be done at BSL-3 laboratory.
- For inactivated samples, testing procedures like RT-PCR and ELISA testing can be done at a BSL-2 laboratory.

BIOSAFETY RECOMMENDATIONS FOR LABORATORIES CONDUCTING DIAGNOSTIC TESTING FOR EBOLA VIRUS DISEASE WITHOUT APPROPRIATE BSL-3/BSL-4 FACILITIES

- Testing procedures like PCR or ELISA for samples should be done in a biosafety cabinet (glove box)-Class III in an isolated laboratory area.

- Once specimens are inactivated, they can be removed from the biosafety cabinet-Class III and the rest of the procedures can be done under BSL-2 conditions.
- All the collected samples can be kept under refrigeration at 2–8 °C up to a week. The biological samples should not be stored under BSL-2 conditions beyond the timeline of necessary shipment.[38]

Thus, the above-documented guidelines recognize existing precautionary measures and biosafety procedures within laboratories that are satisfactory enough to safeguard the laboratory personnel during most testing procedures.[39]

REFERENCES

1. Lyon GM, Mehta AK, Varkey JB, Brantly K, Plyler L, McElroy AK, et al. Clinical care of two patients with Ebola virus disease in the United States. *N Engl J Med* 2014.
2. control cfd. *Ebola virus disease information for Clinicians in U.S. Healthcare settings.* 2014. [cited 12/17/2014]. Available from: http://www.cdc.gov/vhf/ebola/hcp/clinician-information-us-healthcare-settings.html.
3. Mahanty S, Bray M. Pathogenesis of filoviral haemorrhagic fevers. *Lancet Infect Dis* 2004;**4**(8):487–98.
4. Bray M, Geisbert TW. Ebola virus: the role of macrophages and dendritic cells in the pathogenesis of Ebola hemorrhagic fever. *Int J Biochem Cell Biol* 2005;**37**(8):1560–6.
5. Organization WH. *Specific infectious diseases involving potential health risks for travellers.* 2014. [cited 12/17/2014]. Available from: http://www.who.int/ith/diseases/en/.
6. Peters CJ, Jahrling PB, Khan AS. Patients infected with high-hazard viruses: scientific basis for infection control. *Arch Virol Suppl* 1996;**11**:141–68.
7. Roels TH, Bloom AS, Buffington J, Muhungu GL, MacKenzie WR, Khan AS, et al. Ebola hemorrhagic fever, Kikwit, Democratic Republic of the Congo, 1995: risk factors for patients without a reported exposure. *J Infect Dis* 1999;**179**(Suppl. 1):S92–7.
8. Qureshi AI, Chughtai M, Bah EI, Barry M, Beavogui K, Loua TO, et al. High survival rates and associated factors among Ebola virus disease patients at Donka National Hospital, Conakry, Guinea. *J Vasc Interv Neurol* 2015.
9. Schieffelin JS, Shaffer JG, Goba A, Gbakie M, Gire SK, Colubri A, et al. Clinical illness and outcomes in patients with Ebola in Sierra Leone. *N Engl J Med* 2014.
10. Khan AS, Tshioko FK, Heymann DL, Le Guenno B, Nabeth P, Kerstiens B, et al. The reemergence of Ebola hemorrhagic fever, Democratic Republic of the Congo, 1995. Commission de Lutte contre les Epidemies a Kikwit. *J Infect Dis* 1999;**179**(Suppl. 1):S76–86.
11. Bwaka MA, Bonnet MJ, Calain P, Colebunders R, De Roo A, Guimard Y, et al. Ebola hemorrhagic fever in Kikwit, Democratic Republic of the Congo: clinical observations in 103 patients. *J Infect Dis* 1999;**179**(Suppl. 1):S1–7.
12. Formenty P, Hatz C, Le Guenno B, Stoll A, Rogenmoser P, Widmer A. Human infection due to Ebola virus, subtype Cote d'Ivoire: clinical and biologic presentation. *J Infect Dis* 1999;**179**(Suppl. 1):S48–53.
13. Martini GA. Marburg agent disease: in man. *Trans R Soc Trop Med Hyg* 1969;**63**(3):295–302.
14. Kortepeter MG, Bausch DG, Bray M. Basic clinical and laboratory features of filoviral hemorrhagic fever. *J Infect Dis* 2011;**204**(Suppl. 3):S810–6.
15. Locsin RC, Matua AG. The lived experience of waiting-to-know: Ebola at Mbarara, Uganda–hoping for life, anticipating death. *J Adv Nurs* 2002;**37**(2):173–81.

16. control cfd. *Algorithm for evaluation of the returned traveler.* 2014. [cited 12/17/2014]. Available from: http://www.cdc.gov/vhf/ebola/pdf/ebola-algorithm.pdf.

17. control cfd. *Travel and transport risk assessment: guidance for public health authorities and the transport sector.* 2014. [cited 12/17/2014]. Available from: http://www.who.int/csr/resources/publications/ebola/travel-guidance/en/.

18. Organization WH. *Ebola and Marburg virus disease epidemics: preparedness, alert, control, and evaluation.* 2014. [cited 12/17/2014]. Available from: http://apps.who.int/iris/bitstream/10665/130160/1/WHO_HSE_PED_CED_2014.05_eng.pdf?ua=1.

19. Organization WH. *Case definition recommendations for Ebola or Marburg virus diseases.* 2014. [cited 12/17/2014]. Available from: http://who.int/csr/resources/publications/ebola/ebola-case-definition-contact-en.pdf.

20. control cfd. *Guidance for monitoring and movement of persons with potential Ebola virus exposure.* 2014. [cited 12/17/2014]. Available from: http://www.cdc.gov/vhf/ebola/exposure/monitoring-and-movement-of-persons-with-exposure.html.

21. control cfd. *Specimen collection, transport, testing, and submission for patients with suspected infection with Ebola virus disease.* 2014. [cited 12/17/2014]. Available from: http://www.cdc.gov/vhf/ebola/pdf/ebola-lab-guidance.pdf.

22. Gire SK, Goba A, Andersen KG, Sealfon RS, Park DJ, Kanneh L, et al. Genomic surveillance elucidates Ebola virus origin and transmission during the 2014 outbreak. *Science* 2014;**345**(6202):1369–72.

23. Feldmann H. Ebola–a growing threat? *N Engl J Med* 2014;**371**(15):1375–8.

24. Organization WH. *Laboratory guidance for the diagnosis of Ebola virus disease interim recommendations.* 2014. [cited 12/17/2014]. Available from: http://apps.who.int/iris/bitstream/10665/134009/1/WHO_EVD_GUIDANCE_LAB_14.1_eng.pdf.

25. control cfd. *Early recognition is critical for infection control.* 2014. [cited 12/17/2014]. Available from: http://www.cdc.gov/vhf/ebola/pdf/could-it-be-ebola.pdf.

26. Sadek RF, Khan AS, Stevens G, Peters CJ, Ksiazek TG. Ebola hemorrhagic fever, Democratic Republic of the Congo, 1995: determinants of survival. *J Infect Dis* 1999;**179**(Suppl. 1):S24–7.

27. Sanchez A, Lukwiya M, Bausch D, Mahanty S, Sanchez AJ, Wagoner KD, et al. Analysis of human peripheral blood samples from fatal and nonfatal cases of Ebola (Sudan) hemorrhagic fever: cellular responses, virus load, and nitric oxide levels. *J Virol* 2004;**78**(19):10370–7.

28. Ksiazek TG, Rollin PE, Williams AJ, Bressler DS, Martin ML, Swanepoel R, et al. Clinical virology of Ebola hemorrhagic fever (EHF): virus, virus antigen, and IgG and IgM antibody findings among EHF patients in Kikwit, Democratic Republic of the Congo, 1995. *J Infect Dis* 1999;**179**(Suppl. 1):S177–87.

29. Bausch DG, Towner JS, Dowell SF, Kaducu F, Lukwiya M, Sanchez A, et al. Assessment of the risk of Ebola virus transmission from bodily fluids and fomites. *J Infect Dis* 2007;**196**(Suppl. 2):S142–7.

30. Rowe AK, Bertolli J, Khan AS, Mukunu R, Muyembe-Tamfum JJ, Bressler D, et al. Clinical, virologic, and immunologic follow-up of convalescent Ebola hemorrhagic fever patients and their household contacts, Kikwit, Democratic Republic of the Congo. Commission de Lutte contre les Epidemies a Kikwit. *J Infect Dis* 1999;**179**(Suppl. 1):S28–35.

31. Slenczka WG. The Marburg virus outbreak of 1967 and subsequent episodes. *Curr Top Microbiol Immunol* 1999;**235**:49–75.

32. Woo JH, Park HS. Successful treatment of severe sympathetically maintained pain following anterior spine surgery. *J Korean Neurosurg Soc* 2014;**56**(1):66–70.

33. Health MDo. Personal protective equipment (PPE) for infection control: Minnesota Department of Health; [cited 2015]. Available from: http://www.health.state.mn.us/divs/idepc/dtopics/infectioncontrol/ppe/.

34. Prevention CfDCa. Guidance for the selection and use of personal protective equipment (PPE) in healthcare settings: Centers for Disease Control and Prevention [cited 2015]. Available from: http://www.cdc.gov/HAI/pdfs/ppe/PPEslides6-29-04.pdf.

35. Agency USEP. Personal protective equipment: United States Environmental Protection Agency [cited 2015]. Available from: http://www2.epa.gov/emergency-response/personal-protective-equipment.

36. Prevention CfDCa. *Guidance on personal protective equipment to be used by healthcare workers during management of patients with Ebola virus disease in U.S. Hospitals, including procedures for putting on (donning) and removing (doffing)*. Centers for Disease Control and Prevention; April 5, 2015. Available from: http://www.cdc.gov/vhf/ebola/healthcare-us/ppe/guidance.html.

37. Health NYSDo. Revised NYS/NYC laboratory guidelines for handling specimens from patients with suspected or confirmed Ebola virus disease: New York State Department of Health [cited 2015]. Available from: http://www.health.ny.gov/diseases/communicable/ebola/docs/lab_guidelines.pdf.

38. Organization WH. *Algorithm for handling of samples from suspected Ebola virus disease*. World Health Organization; April 5, 2015. Available from: http://www.paho.org/hq/index.php?option=com_docman&task=doc_view&Itemid=270&gid=27912&lang=en.

39. Health BCMo. Draft laboratory biosafety considerations for handling specimens from patients designated as "Suspected Ebola". British Columbia Ministry of Health. Available from: http://www.health.gov.bc.ca/pho/pdf/bc-vhf-laboratory-biosafety-considerations-aug-19-2014.pdf.

40. Association ABS. Risk group classification for infectious agents: American Biological Safety Association. Available from: http://www.absa.org/riskgroups/.

41. McLeod V. *Biosafety levels 1, 2, 3 & 4: Lab Manager*. 2010. Available from: http://www.labmanager.com/lab-health-and-safety/2010/12/biosafety-levels-1-2-3-4?fw1pk=2-.VTXn-fK2eDRY.

42. Organization WH. *Laboratory diagnosis of Ebola virus disease*. World Health Organization; 2014. Available from: http://apps.who.int/iris/bitstream/10665/134009/1/WHO_EVD_GUIDANCE_LAB_14.1_eng.pdf.

43. Canada PHAo. *Interim biosafety guidelines for laboratories handling specimens from patients under investigation for Ebola virus disease*. Public Health Agency of Canada; 2014. Available from: http://www.phac-aspc.gc.ca/id-mi/vhf-fvh/ebola-biosafety-biosecurite-eng.php.

44. *Ebola fever—brief instructions for handling and transport of samples from suspected cases and exposed contacts, including referral for diagnostic confirmation*. Quality Assurance Exercises and Networking on the Detection of Highly Infectious Pathogens (QUANDHIP); 2014. [updated September 8, 2014]. 1.5.4. Available from: http://www.quandhip.info/Quandhip/EN/Home/Ebola_fever_instructions.pdf?__blob=publicationFile.

45. Prevention CfDCa. *Guidance from U.S. Laboratories for managing and testing routine clinical specimens when there is a concern about Ebola virus disease*. Centers for Disease Control and Prevention; 2014. Available from: http://www.cdc.gov/vhf/ebola/healthcare-us/laboratories/safe-specimen-management.html.

Chapter 10

Ebola Virus Disease: Psychological and Social Aspects

Chapter Outline

As I waited for the signs and symptoms of Ebola to develop, I could not sleep. I was very worried about leaving my children as orphans. I knew that once I got Ebola, it meant I would die. So I just sat there, almost too paralyzed to do anything, as I waited for the outcome. I felt defeated. I had nothing to do. This feeling of helplessness continued for some time. I would think all the time.

Data Display 6.4.3: Living under the Threat of Ebola: A Phenomenological Study.

Such was the gravity of the epidemic that it has not only affected the lives of the people physically, but has also caused a significant impact on the psyche of the affected individuals and communities. This chapter highlights the psychological and social implications of the current epidemic and widespread fear, panic, hysteria, and challenges faced by the people affected by Ebola virus disease, the survivors, their families, and the health-care workers. The United Nations (UN) Security Council has declared Ebola virus disease a global threat.[1] The affected nations from the Ebola virus disease outbreak have been branded as "infectious countries." This has led to decline in the economy of these countries. Their agricultural production and trade have become stagnant; thus, creating a large magnitude of economic losses along with fear aggravating the medical consequences of the disease itself.

THE PSYCHOLOGICAL STRESS AND RESPONSE

Psychological stress and manifestations in acute period

Even in the presence of best health care, Ebola virus disease victims were vulnerable to extreme fear, which was most likely due to the dreaded experience. The magnitude of terror, fear, and anxiety felt by the Ebola virus disease patients upon hearing the news of their diagnosis was far greater in comparison to that of human immunodeficiency virus (HIV)[2] infection. When caregivers witnessed the extreme pain and suffering of their patients, it exacerbated the fear of these caregivers becoming infected by the Ebola virus disease.[3] Wendo[4] argues that such anguish might have been due to mass media's portrayal of the disease, which was further fueled by the existing social and cultural stigmas. This fear and panic led to the stigmatization and ostracism of Ebola virus disease patients, their families, survivors, and health workers. Two important aspects that are frequently exaggerated include contagiousness and mortality rate of the disease, which, in the minds of nonmedical and medical personnel alike, approached 100%.

Symptoms in 70 consecutive patients with Ebola virus disease admitted to Donka National Hospital, Conakry, Guinea, included arthralgias (12.86%), myalgias (14.29%), headache (35%), vomiting (50%), fever (46%), and diarrhea (42.86%).[5] These symptoms can be related to underlying psychiatric and psychological processes[3,6–8] and are commonly exacerbated by such underlying processes.[9–11]

The Ebola virus disease has similarities to HIV/AIDS (acquired immune deficiency syndrome)-related stigma which led to avoidance of health care, reduced adherence to antiretroviral medications, increased HIV symptomatology, and the emergence of mental health pathologies including depression.[12–14] This behavior has not only led to shattered familial, social and economic relationships, and infrastructures, but has also created vast barriers for implementation of access, prevention, and treatment strategies (Figure 10.1).[15–20]

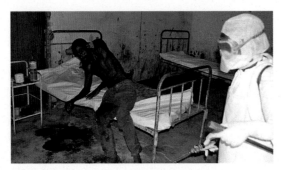

FIGURE 10.1   A Red Cross worker disinfects the hospital room of an Ebola virus disease-affected patient. *Source: AFP.*

Psychological stress and manifestations in postinfectious period

An interesting psychological aspect among Ebola virus disease survivors is the manner in which they responded to hearing the initial news of being declared disease-free. Due to constant suffering caused by physical symptoms of the illness, some expressed distrust of the news disease survival. The label of "survivor" was received with cautiousness and hesitation.[3]

A study conducted by Matua and Locsin[21] in Gulu District, Uganda, found the survivors had "disbelief in their survival" when declared disease-free. Even Kibale[22] participants considered their survival as "impossible" given the fact that no specific medicine had been given to the patients. Even during the survey conducted among survivors of 1995 Kikwit Ebola epidemic by Dee Roo, Ado, Rose, Guimard, Fonck, and Colebunders[23] found a similar incidence of "denial" by those who survived.

People exposed to life-threatening illness such as Ebola virus disease, HIV, or cancer often suffer from several negative psychological symptoms often classified under posttraumatic stress disorder (PTSD). This mental trauma leads to severe psychiatric problems in affected individuals. According to Van der Kolk et al.,[24] when individuals recovering from a traumatic event fail to transcend their negative experience, it could be an indication of a more severe psychological problem, identified as PTSD. Kibale survivors, Dee Roo et al.,[22,23] have reported the psychological effect of the survivor during 1995 Kikwit outbreak, where there was a widespread fear, discrimination, shame, and stigmatization observed.

Whenever I am reminded about Ebola, I find I cannot control crying because recalling my situation stresses me. And when I reflect about my experience, sometimes I think a lot because the illness didn't leave me the way I was before. The illness has left me with many weaknesses…

Data Display 6.3.3.; Living under the Threat of Ebola: A Phenomenological Study.

A study on survivors from the Ebola virus disease outbreak in Mbarara, Uganda, found that contacts would experience unusual headaches and a feverish sensation when simply asked about their Ebola virus disease experience. It appears, therefore, psychosomatic disturbances may persist long after being infection-free.[3]

A survey among 34 survivors from Ebola virus disease survivors in Kikwit General Hospital in 1995 outbreak focused on psychological symptoms. In summary, the study defines experiences of the 1995 Ebola epidemic survivors in Kikwit, Democratic Republic of Congo. Before falling ill, most survivors took care of a sick family member and none of them had ever heard of Ebola virus disease before they developed symptoms. This is part of the reason why they never suspected that they would get infected by the virus. In the beginning of the Ebola virus disease outbreak, they experienced fear, denial, and shame. Their belief in God helped them through the phase of suffering. Their most horrible experience was seeing their own people dying in the isolation ward of Kikwit

Hospital along with the hesitancy of health-care workers to treat the affected. Therefore, more attention is needed to address the psychological consequences during Ebola virus disease epidemic.[23]

There was a study conducted on 105 Ebola virus disease survivors when discharged from Ebola isolation unit at Donka National Hospital described "post-Ebola virus disease syndrome" that was a combination of physical and psychological symptoms. In those who were in subacute period of recovery reported symptoms of anorexia, chronic joint, and muscle pains along with several other symptoms like difficulty in short-term memory, headaches, sleep difficulties, insomnia, dizziness, abdominal pain, constipation, decreased exercise tolerance, decreased libido, and sexual dysfunction. The most noticeable psychiatric symptoms seen among them were the change in the level of self-confidence due to constant exposure to rejection and ostracism within the community.[25]

HEALTH-CARE RESPONSE

This psychological stress was further augmented by the absence of a structured psychological support system and the lack of well-trained mental health professionals to cope with these challenges. A classic example would be that, in Liberia and Sierra Leone, each of them has only one practicing psychiatrist, more than a few dozen mental health nurses, and about 100 trained paramedical professionals for the management of common mental disorders.[26] Hence, it is very important to assess the situation and identify the necessary measures needed to cope with those challenges.

First, there is a need for rapid assessment of outbreak-related mental stressors within the general public and related health workers, which can be done by using trauma signature analysis, an evidence-based method that helps in analyzing populations exposed to an extreme event, thus providing an actionable pathway for highly targeted support programs.[27] The technique of trauma signature analysis was used in a major disaster like 2011 Japan tsunami, where the features of the event was identified, a hazard profile was created, and the severity of the exposure and related mental risk factors was assessed.[28]

Next, in cases of emergency situations like the Ebola virus disease outbreak, an Inter-Agency Standing Committee (IASC) guidelines on mental health and psychological support should be adopted.[29] These guidelines provided support in extreme situations like Syrian war refugees, Philippine typhoon survivors, and Nepalese survivors of political violence. The main features of IASC guidelines are (1) to restore security and basic services to the people affected, (2) to consolidate family and community networks, (3) to provide support to distressed individuals, and (4) to offer specialized mental health intervention for severely affected survivors. Lastly, the target populations for the intervention should include stigmatized survivors, mourning family members, ostracized orphans, and related health-care workers.[30]

ESCALATION BY MEDIA

The media has also played a major role in escalating the fear among the general population under the guise of sensationalism. The front page of the newspaper headlines stating "Ebola claiming Gulu Hospital head" (IRIN News, December 6, 2000) and "The disease of year 2000 was the Ebola fever" (New Vision, January 8, 2001) are examples of sensationalism prior to the current outbreak (Figure 10.2).

In a workshop symposium titled "Disease, Communication, and the Ethics of (In) Visibility," participants identified instances where historical representations of certain medical conditions raise questions about the ethics of spectatorship. The workshop also focused on the tribulations that arose from the ongoing necessity to evoke empathy in the representation of individuals' suffering while striving not to perpetuate social and economic inequalities. Finally, the lecture also considered the intersection of visibility with the complex nexus of health, sexuality, and global/national politics. Such considerations highlight the relationship between the necessity of making diseases visible, the mechanisms of legal and visual censorship, and the overall ethics of viewing and spectatorship.[31]

RESPONSE OF OFFICIALS AND SPOKE PERSONS, AND PUBLIC RITUALS

Another important factor that had influenced this epidemic is the manner in which the officials handled the situation. Accordingly, the amount of information and misinformation regarding the Ebola virus disease deeply impacted the psychology of the nation (West African countries).[32] According to Meredith Stakem, a health and nutrition advisor for Catholic Relief Services in West Africa, one of the lesser known psychological factors that has triggered an even greater reaction among the ignorant community is the fact that health workers respond to the Ebola virus disease by wearing head-to-toe protective gear that looks like "space suits."

FIGURE 10.2 A man reads a local newspaper with headline "Ebola Virus Kills Liberian in Lagos," Saturday, July 26, 2014. *AP Photo/Sunday Alamba.*

The previous largest Ebola virus disease outbreak was first reported in the year 2000 in the surroundings of Lacor Hospital in the northern Ugandan town of Gulu.[33] A Medical Superintendent at Gulu Hospital, Uganda, described the situation as "very wild." Following this statement 4 days later, the mourners at a funeral of an Ebola virus-affected person stripped themselves naked and left their clothes at the funeral, fearing that they might carry virus to their respective homes. In response to this, the Ministry of Health responded, "sometimes the response is extreme, but it shows that people are taking the message." This response was taken to extremes by the local people, who started to drink and bath in "Jik" bleach (a substance promoted during the epidemic as an effective disinfectant against the virus) "in a desperate attempt to rid themselves of Ebola."[34]

The public display of stigmatization increased substantially within the community after outbreak, with one of the affected villagers of Gulu town, Uganda, described "people eyeing each other suspiciously, suspecting everybody else to be a carrier of the killer virus." On the 19th day of the outbreak, there was a traditional Acholi ritual performed in which the people of Gulu "erupted in wild shouting, jumping, and running," which was intended to "exorcise" their land of the virus.[34] Such activities promote stigmatization not just to the affected individuals, but also to the survivors. McEwen[10] as cited in Major and O'Brien[11] agrees that people who experience stigmatization and ostracism have a greater risk of developing mental and physical health problems in comparison to non-stigmatized populations.

THE HEALTH-CARE WORKERS AND CAREGIVERS

Even though health-care workers have in-depth knowledge regarding the disease and well-developed protocols to ensure safety, the magnitude of exposure and direct visualization of the disease manifestations pose unique challenges. Health-care workers struggle with the sense of responsibility while constantly facing the fear of becoming victims themselves. A good example is the story of a young male health-care worker who contracted Ebola virus disease while taking care of a patient and was admitted to the isolation ward of the infectious disease hospital in Lagos, Nigeria. Three weeks following the admission, he appeared confused and was sleeping poorly, talking vaguely, and displaying unruly, disruptive behavior toward those taking caring of him. After a mental status examination, the health-care worker was found to be restless, agitated, and fearful about the outcome of his illness. His attention and concentration were affected. A diagnosis of adjustment disorder with mixed disturbances of emotion and conduct was made.[35] Reports of other health-care workers acquiring infection during performance of duties create constant psychological stress for workers and families. Dee Roo et al.[23] reported that witnessing the death of friends and colleagues in the isolation unit was the most negative psychological feeling experienced by medical staff.

FIGURE 10.3   Health-care workers wearing protective gear at Medecins Sans Frontieres' (MSF) Ebola virus disease treatment center. *Source: Kjell Gunnar Beraas/MSF/AP.*

Nonmedical personnel, such as family members in West Africa, might also provide a large part of Ebola virus disease victim care. These caregivers reported that witnessing the death of their loved ones was one of the worst psychological experience leading to sleeplessness, anxiety, and despair.[3] Caregivers also found it hurtful to have persistent memories of the deceased relatives, which continued to linger in their minds. Overall, the Ebola virus disease has affected the community in different ways, which include social stigma, social withdrawal, lingering negative experience, obliteration of sufferers' actualities and possibilities, and the reality that nearly everyone is at risk for infection (Figure 10.3).[3]

Acceptance into the society

Thanks be to God, I am cured. But now I have a new disease: the stigmatization that I am a victim of. I should have died. Now, everywhere in my neighborhood, all the looks bore into me like I'm the plague

The previous statement was from an Ebola virus disease survivor who was not just a regular survivor, but a doctor from Guinea who had beaten the odds and survived Ebola virus disease and now faces a second disease called "Stigma" that still lingers even after he completely recovered from the physical effects of a deadly disease.[36] This extreme level of stigmatization and ostracism toward the Ebola virus disease-affected individuals, survivors, their families, caregivers, and related health workers is based on the fear that has gripped the affected regions and exacerbated by suspicion and rumors.

Many of our neighbors won't come to our house now… My friends don't visit, thinking that if they come near our house, they'll catch the virus.

Survivor Fudia Sesay of Sierra Leone.

During Kikwit outbreak in 1995, the 190 Ebola virus disease survivors who returned home found that the people from the community had burned their clothes, beds, and even their huts. In some instances, their family members had fled and people even refused to trade with them.[34] The fear among the people in the community was so great that, even children suspected of having Ebola virus disease, when found to be seronegative or infection-free, were not adopted by their own families out of fear of getting infected.[37] In another scenario in Sierra Leone during the Kikwit outbreak, Ebola virus disease survivors were branded as "witches" for surviving. Even after emerging from what can be described as a "glimpse of hell," these survivors could not find an easy life on the other side of the Ebola virus disease ward.[38]

Another important aspect of behavior among the community is the way a suspected Ebola virus disease patient was handled. Mason[39] described observations from 2007 Ebola Bundibugyo outbreak, where Ebola virus disease patients and their families were forcibly quarantined by the community to an extent where the members of labeled families were stoned if they try to leave their homes, for the fear of spreading Ebola virus disease. While, on the one hand, the epidemic control protocols succeeded in minimizing the spread of the infection, on the other hand, the dehumanizing actions, especially the manner in which the suspected persons were handled, increased the panic in the public as they started to believe the possibility that they might also be humiliated in the same way. The spread of infection, fear, and stigma among the general population has also increased to an extent where the attendance in church and mosques tumbled. The clergy abandoned age-old rituals involving direct body contact that, although it helped to minimize the spread of infection, increased the anxiety levels and fear of divine abandonment.

The stigmatization and ostracism in society follow from cultural beliefs and practices that are heavily influenced by the past. McLaughlin and Braun[40] stated that cultural values influence patient's attitudes about obtaining and seeking treatment information. The individual's selection of care, including the conduct and reaction to the disease or illness, is influenced by the cultural practices and beliefs the individual follows.[3] Interestingly, a very small percentage of children are affected by Ebola virus disease. This low number could be due to cultural practices in which children are kept away from sick family members in regions with long-standing histories of infectious disease outbreaks, thereby reducing the transmission of such infections to children.[41,42]

The social stigma around the disease has caused the further spread of infection because of false information provided by those who may have been exposed, because "people don't want to be associated with the health crisis. When you ask them about their travel history they give you false information" (Ebola Survivor, Sulaiman Kanneh Saidu, Health worker, Sierra Leone, who caught the disease during the examination of a sick woman who lied about her contact with the disease).[43] Barret and Brown[44] reported that ignorance and lack of understanding of a disease could contribute significantly toward escalating fear, stigma, and ostracism. A classic example is the incidence of "ignorance-induced fear and panic" that occurred in the city of Surat, Western part of India, where in 1994 the "pneumonic plague" epidemic resulted in over a half million people fleeing from the city.[44–46]

INTERACTION OF HEALTH-CARE WORKERS WITH SOCIETY

Even the caregivers reported negative experiences associated with caring for close relatives and friends diagnosed with Ebola virus disease, characterized by social isolation, rejection, and discrimination. These experiences were similar to those experienced by survivors. The fear of getting infected, high rates of Ebola virus disease-related deaths, and extreme caution used by the epidemic response team increased the antisocial incidents directed toward Ebola virus disease patients, along with their caretakers and family members. Even during the epidemic in 2000 and 2001, several instances of harassment and abandonment were observed.[47]

VIOLENT BEHAVIORS WITHIN SOCIETY

This stigmatization has reached to such an extent where some even believe the ongoing epidemic is due to "divine retribution" for past sins. In April 2014, a rumor spread that foreigners were spreading the disease led to an attack on a Doctors Without Borders (MSF) clinic in rural Guinea, forcing them to abandon its mission. Even in Sierra Leone, few local leaders spread rumors that "the white people" were conducting experiments that had caused the disease.[48]

There have been several more instances reported like the incidences reported in Butagira's[49] article "Bundibugyo Residents Shun Ebola Survivors" (Daily Monitor Newspaper, December 21, 2007) and Mugisa's article "Ebola isolation center closed over violence" (New Vision, December 9, 2007). Thawite, Ssengendo, Etengu, and Kagiri[50] describe villagers vandalizing an Ebola virus disease isolation unit; this fear-driven "mob action" was viewed as a necessary action to protect the community from the "infection source" (i.e., the patients).

There was also anger among the public about how Ebola virus disease patient burials were handled. Their loved ones who had died from the disease were not buried according to their usual traditions. Their typical tradition of burial includes bathing the dead person and then burying them which involves direct contact. Because of the outbreak, the government officials have taken up the job of burial through trained health workers. Although such precautions were necessary to any direct contact with the people affected, the perception among families was different. In one district of Guinea, police fired tear gas at a mob that was trying to raid the morgue in order to give their loved ones proper burials despite the risk of contagion (Figures 10.4–10.7).[48]

SOCIAL STIGMA IN THE UNITED STATES

It is strange that this fear, stigmatization, and ostracism are not limited to underdeveloped countries, but are also seen in developed nations like the United States. During intensified news coverage in October 2014 in relation to cases of

FIGURE 10.4   Encouraging message to bring awareness in the community in Liberia. *Source: Ahmed Jallanzo/EPA; The Irish Times.*

FIGURE 10.5   The residence of Ebola virus disease survivor burnt during the 2003 DRC Ebola outbreak. *Source: Ebola in Congo: virus, poacher and fetish, Mbomo © Epelboin 2004.*

FIGURE 10.6   Ebola virus disease survivors tell their stories at the Survivors Conference in Kenema, Sierra Leone-WHO, October 2014. *Source: WHO/S.Gborie.*

Ebola virus disease in the United States, circumstantial observations suggested that many Americans were found to be anxious about the Ebola virus disease. Given the insignificant risk of infection, their apprehension was perhaps driven by apparent rather than actual risk.[51]

> *If I'm on the metro, I don't talk. If I'm on the bus, I don't talk. If people hear the accent, they think you are Liberian, Then you have Ebola.*

FIGURE 10.7   A health worker sprays the body of a man thought to have died of Ebola virus disease in Monrovia, Liberia. *Source: AP Photo/Abbas Dulleh.*

This was the response of a Liberian-American Charles Roberts when speaking to ABC News. There have also been reports of Liberians in Washington being instructed to temporarily leave work. This situation was seen all over the country where Liberians experienced stigmatization just because they were closely associated with a country currently facing an Ebola virus disease outbreak.[52]

Fear in a country of 316 million people with less than 10 patients who have actually contracted the disease was driven by distrust, leading to panic, discrimination, and even violence. People began to doubt experts, criticize health officials for any perceived mistake, and wonder why more attention was not being paid to prepare for worst-case scenarios. This distrust extended beyond the health-care system when people became suspicious about their safety in the workplace, worrying about the protection of their children at schools, and looking sideways at the person next to them on the subway, on the bus, or in neighboring houses.

To alleviate this tension, the best approach requires government officials to join with respected health-care leaders and organizations to address the public.[53] To curb negative feelings among the general public, a large number of awareness campaigns have been conducted, including on social media like Twitter, where tweeted pictures of Liberians were presented holding up signboards (Figure 10.8).

CONCLUSION

One of the biggest challenges extends beyond finding a better way to not only stop the Ebola virus disease transmission, but also stop the stigmatization and ostracism that rapidly develops within the community. Hence, it remains critical to understand and anticipate the societal reaction to the Ebola virus disease and other pandemic outbreaks. Health-care organizations must implement strategies in order to prevent and resolve such occurrences within the society (Figure 10.9).

FIGURE 10.8   People using social media to fight against stigmatization using the hashtag campaign #IamALiberianNotAVirus. *Source: Shoana Cachelle/Cachelle Photography.*

FIGURE 10.9   A car drives past a public health advertisement against the Ebola virus disease in Monrovia, October 8, 2014 (Reuters/James Giahyue).

Dissemination of accurate information to communities affected by the Ebola virus disease can be done by educating and training highly respected community leaders who have ability to influence the community to fight the stigma.[2] Multilevel community involvement that includes accurate information about disease, counseling, and testimonials from persons who have survived the disease, along with caregiver stories, can help decrease the stigma in the society (Figure 10.10).[54]

Collective strategies, including empathy-based contact that promotes programs dedicated to skill building, training local health workers, distributing pamphlets, door-to-door campaigns, and visiting religious places and religious leaders can be beneficial in reducing the fear and stigma within communities.[55–57]

FIGURE 10.10 Health-care workers demonstrating on the streets to bring in awareness in the community. *Source: CNN.*

Even in the United States, there is a need for a science-based risk communication strategy to counter the tendency of the media to exacerbate fear felt by the public. According to researchers, even indirect exposure to events like September 11 and Boston Marathon Bombing through the media outlets is sufficient enough to generate anxiety and elevate risks for common psychological disorders.[58] Therefore, it is important to provide precise and reassuring public health message.[30]

Pamphlets being distributed to the general public as a part of Ebola virus disease awareness campaign. *(Source: AP/BBC News.)*

Another powerful method to fight the stigma is the "way the health-care workers treat a discharged patient" said Corinne Benazech, representative of Doctors without Borders (MSF) in Guinea. "The patient never leaves alone" when discharged; all of the health-care workers individually shake hands with the survivor. The discharged patients then receive a certificate that they are completely cured (Figure 10.11).[36]

The way country leaders treat survivors also affects the attitudes of the community. For example, at a global level, President Obama's public hug with Ebola virus disease survivor Nina Pham was symbolic of tackling the stigma within the community, whereas President Koroma of Sierra Leone avoided physical contact with the survivors he met.[59] The international response is hence very crucial in mobilizing the people to change their outlook toward the dreadful disease. In one statement, the Liberian president Elle Johnson Sirleaf said when American doctors were sent to Liberia, "You can't imagine the difference it made in the hopes of the people. It inspired them to do more."

The role of the print media in presenting the situation to the public also plays a major role. Emphasizing the positive aspects of the situation will bring in a sense of hope and may help in changing the attitude of the people, which can eventually calm down the excessive negative reactions within the community. Researchers recommended that entertainment programs which capture the mind are particularly effective as people are more absorbed in a play are more open to both receiving and to consider a change in the attitude and behavior.

FIGURE 10.11 Ebola virus disease survivor holding a discharge certificate at Medecins Sans Frontieres' (MSF) treatment center. *Source: Martin Zingg/MSF.*

The viewers sometimes develop a personal bond with the imaginary characters in programs. This occurrence is called "parasocial interaction." Therefore, combining entertainment education in radio, television, and social networking sites can create a greater impact among the population.[60]

There was a campaign started by UN wherein they launched a smartphone application called #ISurvivedEbola. Survivors in Guinea, Sierra Leone, and Liberia were given smartphones so that they could share their stories, in effect helping to fight the existing stigma. There is a study done by Pew Research Center (The George Washington University School of Media and Public Affairs) revealing a relationship between electronic news media publishing and Twitter activity around significant events such as the Ebola virus disease. The influence of electronic news and social media should be used to the advantage of healthcare organizations to develop social media campaigns that can create a positive impact on the general public.

In a random sampling of tweets, it was observed that there was a high frequency of Ebola virus disease-related tweets in association with negative emotions, nervousness, rage, swearing, and discrepant thinking[61] in comparison to influenza. These Twitter data can provide public health practitioners with a measurable indication of anxiety, anger, or negative emotions in the general public. This indicator could help to lessen nervousness and correctly communicate facts about Ebola virus disease risk.[51]

Above attempts to slow the transmission of the virus by providing soaps and chlorine households and training local authorities to take up the responsibility of the burial process of the victims could be of great help to disease prevention. The government could provide better health-care facilities to hospitalized Ebola virus disease patients, early detection of the cases, a plan to handle Ebola virus disease orphans and survivors, and psychiatric counseling to survivors dealing with PTSD and mental trauma due to stigmatization (Figure 10.12).

The involvement of anthropologists to build a new collective platform (like the Ebola Response Anthropology Platform) and networks (like the Ebola Anthropology Initiative discussion board and Listerv) will help Ebola virus disease research and intervention by providing data analysis capabilities,

FIGURE 10.12   Medecins Sans Frontieres' (MSF) 1000th survivor—Ebola virus disease survivor Kollie James and his father Alexander, MSF health promotion officer in Foya, at the guest house in Foya after his discharge. *Source: Katy Athersuch/Courtesy of Doctors without Borders.*

organizing academic institutional resources, sharing actionable recommendations, providing expertise and technical information, and collaborating on research initiatives that will eventually help in avoiding wasteful replication of effort.[62,63] There is also a strong need for integration of social sciences during global health crisis that can rapidly mobilize all experts during Ebola virus disease response and can bring relevant local contextual, medical, epidemiological, and political information on global health emergencies.[63,64]

There is also a need to learn lesson from AIDS/HIV epidemic when tackling the Ebola virus disease. These epidemics are similar in that they display a mix of scientific ignorance and fear in the public reactions to the return of healthcare workers from the front lines in West Africa. When these individuals return to their home countries from fighting the Ebola virus disease, politicians and media sources magnify these reactions by perpetuating the fear-driven doubt and exclusion of whole classes of people. These responses are similar to the emergence of AIDS/HIV in the 1980s.[65]

The Sierra Leone Ebola Emergency Operations Psychosocial Consortium (consists of members of the Sierra Leone government, nongovernmental organizations, and donor agencies) assessed survivors' health, psychosocial, and financial needs and their interest in supporting the Ebola virus disease response. In October 2014, the consortium assessed survivor needs in three districts (Bo, Kenema, and Bombali). Assessment included (1) convening a National Survivor Conference in the Kenema District where they conducted five focus groups with 36 survivors, (2) conducting in-depth interviews with 12 survivors, (3) conducting five additional district-specific focus groups with a total of 51 survivors, and (4) observing six survivor wellness center counseling sessions. The focus group discussions and in-depth interviews included assessing survivor experiences around postdischarge support, their feelings about surviving the Ebola

virus disease, and specific jobs or tasks survivors could perform. On completion of the evaluation, findings were shared with select district-level Emergency
Operations Center staff and partners involved in the response to improve and
organize survivor services.

This process helped survivors considerably in improving their mental stability and positive attitude and has generated self-advocacy, eventually leading
survivors to share their stories directly with their local communities, with Ebola
virus disease patients, and with a larger audience through radio and other broad-
cast media. They also expressed interest in participating in Ebola virus disease
care (treatment support and direct care), along with providing moral support to
other Ebola virus disease patients to give them hope. Many indicated that supporting themselves with this work would help restore their own dignity.

The Sierra Leone Ebola Emergency Operations Psychosocial Consortium
also is coordinating partners and districts to improve the initial and ongoing
psychosocial support for survivors. A counselor–client flipbook is in development, containing a series of pictures with information to help change health
behaviors, and will serve as an aid for counselors to ensure steady and comprehensive discharge planning and counseling for all survivors throughout Sierra
Leone. Likewise, a comprehensive survivor packet has been designed to ensure
the consistent provision of resources to survivors upon discharge. The packet
includes a mattress, bed sheets, a blanket, a towel, a pillow, a water bucket, a
cell phone, utensils, a cooking pot, laundry soap, bar soap, a toothbrush and
toothpaste, a mosquito net, a set of clean clothes and under garments, plastic
sandals, food, cash, condoms, and multivitamins.

To assist with survivor reintegration in the community, the consortium recommends that counselors accompany survivors when returning to their home
villages to facilitate reunification and reintegration of survivors into their communities. The reintegration process also includes trained counselors speaking
with local traditional authorities and other community members about the survivor's status, the importance of survivor acceptance, and ways in which the
community can support the survivor. In addition, stigma mitigation educational
materials targeting the community have been developed and implemented.[66]

REFERENCES

1. United Nations Security Council. *Resolution 2177. Security Council 7268th 2014*. 2014. Available from: http://www.un.org/en/ga/search/view_doc.asp?symbol=S/RES/2177.
2. Group NCHSPT. Results of the NIMH collaborative HIV/STD prevention trial of a community
popular opinion leader intervention. *J Acquir Immune Defic Syndr* 1999;**1999**(54):204.
3. *Living under the threat of Ebola: a phenomenological study*. 2014. [Cited 28.03.15]. Available from:
http://uir.unisa.ac.za/xmlui/bitstream/handle/10500/13822/thesis_matua_ag.pdf?sequence=1.
4. Wendo C. Italy aids Ebola orphans. *The New Vision* 2001. p. 6.
5. Qureshi AI, Chughtai M, Bah EI, Barry M, Béavogui K, Loua TO, et al. High survival rates and
associated factors among Ebola virus disease patients hospitalized at donka national hospital,
conakry, Guinea. *J Vasc Interv Neurol* 2015;**8**.

6. Locsin RC, Matua AG. The lived experience of waiting-to-know: Ebola at Mbarara, Uganda-hoping for life, anticipating death. *J Adv Nurs* 2002;**37**:173–81.

7. Jacoby A, Snape D, Baker GA. Epilepsy and social identity: the stigma of a chronic neurological disorder. *Lancet Neurol* 2005;**4**:171–8.

8. National Cancer Institute. *Post-traumatic stress disorder*. 2012. Available from: http://www.cancer. gov/cancertropics/pdg/supportivecare/post-traumatic-stress/Patient/page2/AllPages/Print.

9. Cordova MJ, Andrykowski MA. Responses to cancer diagnosis and treatment: posttraumatic stress and posttraumatic growth. *Semin Clin Neuropsychiatry* 2003;**8**:286–96.

10. McEwan BS. The neurobiology of stress: from serendipity to clinical relevance. *Brain Res* 2000;**886**(1–2):172–89.

11. Major B, O'Brien LT. The social psychology of stigma. *Annu Rev Psychol* 2005;**56**(1):393–421.

12. Vanable PA, Carey MP, Blair DC, Littlewood RA. Impact of HIV-related stigma on health behaviors and psychological adjustment among HIV-positive men and women. *AIDS Behav* 2006;**10**(5):473–82.

13. Sevelius JM, Patouhas E, Keatley JG, Johnson MO. Barriers and facilitators to engagement and retention in care among transgender women living with human immunodeficiency virus. *Ann Behav Med* 2014;**47**:5–16.

14. Katz IT, Ryu AE, Onuegbu AG, Psaros C, Weiser SD, Bangsberg DR, et al. Impact of HIV-related stigma on treatment adherence: systematic review and meta. *J Int AIDS Soc* 2013;**2013**(16):18640.

15. Mahajan AP, Sayles JN, Patel VA, Remien RH, Ortiz D, Szekeres G, et al. Stigma in the HIV/AIDS epidemic: a review of the literature and recommendations for the way forward. *AIDS* 2008;**22**:S67.

16. Ekstrand ML, Ramakrishna J, Bharat S, Heylen E. Prevalence and drivers of HIV stigma among health providers in urban India: implications for interventions. *J Int AIDS Soc* 2013;**16**:18717.

17. Rahmati-Najarkolaei F, Niknami S, Aminshokravi F, Bazargan M, Ahmadi F, Hadjizadeh E, et al. Experiences of stigma in healthcare settings among adults living with HIV in the Islamic Republic of Iran. *J Int AIDS Soc* 2010;**13**:27.

18. Pulerwitz J, Michaelis A, Weiss E, Brown L, Mahendra V. Reducing HIV-related stigma: lessons learned from Horizons research and programs. *Public Health Rep* 2010;**125**:272.

19. Mbonu NC, Van den Borne B, De Vries NK. A model for understanding the relationship between stigma and healthcare-seeking behavior among people living with HIV/AIDS in sub-Saharan Africa. *Afr J AIDS Res* 2009;**8**:201–12.

20. Churcher S. Stigma related to HIV and AIDS as a barrier to accessing health care in Thailand: a review of recent literature. *WHO South-East Asia J Publ Health* 2013;**2**:12.

21. Matua GA, Locsin RC. Conquering death from Ebola: living the experience of surviving a life-threatening illness. In: Lee AV, editor. *Coping with disease*. New York: Nova Science; 2005. p. 121–73.

22. Kibale District Local Government. *Kibale district, Republic of Uganda 2012*. 2012. Available from: http://www.kibaale.go.ug/.

23. De Roo A, Ado B, Rose B, Guimard Y, Fonck K, Colebunders R. Survey among survivors of the 1995 Ebola epidemic in Kikwit, Democratic Republic of Congo: their feelings and experiences. *Trop Med Int Health* 1998;**3**(11):883–5.

24. Van der Kolk B, McFarlane AC, Weisaeth L, editors. *Traumatic stress: the effects of overwhelming experience on mind, body and society*. New York: Guilford; 2007.

25. Qureshi AI, Chughtai M, Loua TO, Pe Kolie J, Camara HF, Ishfaq MF, N' Dour CT, Beavogui K. Study of Ebola virus disease survivors in Guinea. *Clin Infect Dis* 2015;**61**(1):1035–42.

26. Levin A. Response to Ebola crisis will require attention to MH needs. *Psychiatr News* 2014;**49**(20):1–26.

27. Shultz JM, Neria Y. Trauma signature analysis: state of the art and evolving future directions. *Disaster Health* 2013;**1**(1):4–8.

28. Shultz JM, Forbes D, Wald D, et al. Trauma signature of the Great East Japan Disaster provides guidance for the psychological consequences of the affected population. *Disaster Med Public Health Prep* 2013;**7**(2):201–14.

29. Committee I-AS. *IASC guidelines on mental health and psychosocial support in emergency settings*. Geneva, Switzerland: Inter-Agency Standing Committee; 2007.

30. Shultz JM, Baingana F, Neria Y. The 2014 Ebola outbreak and mental health: current status and recommended response. *JAMA* 2015;**313**(6):567–8.

31. Pietrzak-Franger MM, Holmes MS. Disease, communication, and the ethics of (in)visibility. *J Bioeth Inq* 2014;**11**(4):441–4. http://dx.doi.org/10.1007/s11673-014-9588-2. Epub 13.11.14.

32. Harlan C. *An epidemic of fear and anxiety hits American amid Ebola outbreak*. 2014. Available from: http://www.washingtonpost.com/buisness/economy/an-epidemic- of-fear-and-anxiety-hits-americans-amid-ebola-outbreak/2014/10/15/0760fb96–54a8-11e4-ba4b-f6333e2c0453_story. html.

33. WHO. WHO: outbreak of Ebola hemorrhagic fever, Uganda, August 2000–January 2001. *Wkly Epidemiol Rec* 2001;**2001**(76):41–6.

34. Kinsman J. "A time of fear": local, national and international responses to a large Ebola outbreak in Uganda. *Glob Health* 2012;**8**(15):2012.

35. Mohammed A, Sheikh TL, Gidado S, Absud-Salam IA, Adeyemi J, Olayinka A, et al. Psychiatric treatment of a health care worker after infection with Ebola virus in Lagos, Nigeria. *Am J Psychiatry* 2015;**172**(3):222–4. http://dx.doi.org/10.1176/appiajp201414121576.

36. Available from: http://www.huffingtonpost.com/2014/04/28/ebola-stigma_n_5226011.html; 2014.

37. Bitchell RE. *After losing parents to Ebola, orphans face stigma*. 2014. Available from: http://www.npr.org/blogs/goatsandsoda/2014/10/03/353473707/after-losing-parents-to-ebola-orphans-face-stigma.

38. WHO. Available from: http://www.who.int/features/2014/post-ebola- syndrome/en/; 2014.

39. Mason C. *Coping with Ebola stigma*. 2008. Available from: http://www.monitor.co.ug/artman/publish/news/copingwithEbolastigma.shtml.

40. McLaughlin LA, Braun KL. Asian and Pacific Islander cultural values: considerations for health care decision-making. *Health Soc Work* 1998;**23**(2):116–26.

41. WHO Ebola Response Team. Ebola virus disease in West Africa: the first 9 months of the epidemic and forward projections. *N Engl J Med* 2014;**371**. http://dx.doi.org/10.1056/NEJM0a1411100.

42. Peacock G, Uyeki TM, Rasmussen SA. Ebola virus disease and children: what pediatric health care professionals need to know. *JAMA Pediatr* December 2014;**168**(12):1087–8. http://dx.doi.org/10.1001/jamapediatrics20142835.

43. News S, 2014; Available from: http://sputniknews.com/world/20140815/192048080.html.

44. Barret R, Brown PJ. Stigma in the time of influenza: social and institutional responses to pandemic emergencies. *J Infect Dis* 2008;**197**:S34–7.

45. Roja R. *The Black death revisited: India's 1994 plague epidemic*. 1977. Available from: http://www.rrojasdatabank.info/uh_b3.htm.

46. Shah G. Public health and urban development: the plague in Surat. In: Barret R, Brown PJ, editors. *Stigma in the time of influenza: social and institutional responses to pandemic emergencies*. 2008. J Infect Dis 1977;**197**(Suppl. 1):S35.

47. Hewlett BS, Amola RP. Cultural contexts of Ebola in northern Uganda. *Emerg Infect Dis* 2003;**9**:1242.
48. CNN. Available from: http://www.cnn.com/2014/07/24/opinion/garrett-ebola/.
49. Butagira T. Bundibugyo residents shun Ebola survivors. *Dly Monit* December 21, 2007:2–4.
50. Mugisa A, Thawite J, Ssengendo A, Etengu N, Kagiri L. Ebola isolation centre closed over violence. *New Vis* December 9, 2007:1–3.
51. Fung IC, Tse ZTH, Cheung CN, Miu AS, Fu KW. Ebola and the social media. *Lancet* 2014;**384**(9961):2207.
52. The Guardian. Available from: http://www.theguardian.com/world/2014/oct22/ebola-liberia-not-virus-stigma.
53. Sharfstein JM. On fear, distrust, and Ebola. *JAMA* 2015;**313**(8):784. http://dx.doi.org/10.1001/jama2015346.
54. Apinundecha C, Laohasiriwong W, Cameron MP, Lim S. A community participation intervention to reduce HIV/AIDS stigma. Nakhon Ratchasima province, northeast Thailand. *AIDS care* 2007;**19**:1157–65.
55. Parker R, Aggleton P. HIV and AIDS-related stigma and discrimination: a conceptual framework and implications for action. *Soc Sci Med* 2003;**57**:13–24.
56. Lebel TP. Perceptions of and responses to stigma. *Soc Compass* 2008;**2**:409–32.
57. Shih M. Positive stigma: examining resilience and empowerment in overcoming stigma. *Ann Am Acad Polit Soc Sci* 2004;**591**:175–85.
58. Neria Y, Sullivan GM. Understanding the mental health effects of indirect exposure to mass trauma through the media. *JAMA* 2011;**306**(12):1374–5.
59. African Research Institute. Available from: http://www.africaresearchinstitute.org/blog/ebola-stigma/.
60. Margaret Miller OJ. *Entertainment media can help change behaviors and stop the Ebola outbreak.* 2014. [Cited 15.04.15]. Available from: http://blogs.worldbank.org/publicsphere/entertainment-media-can-help-change-behaviors-and-stop-ebola-outbreak.
61. Pennebaker JW, Chung CK, Ireland M, Gonzales A, Booth RJ. *The development of psychometric properties of LIWC.* 2007.
62. Janes CR, Corbett KK, Jones JH, Trostle J. Emerging infectious diseases: the role of social sciences. *Lancet* 2012;**380**:1884–6.
63. Abramowitz SA, Bardosh KL, Leach M, Hewlett B, Nichter M, Nguyen VK. *Lancet* 2015;**385**(9965):330. http://dx.doi.org/10.1016/S0140-6736(15)60119-2.
64. Institute of Medicine and National Research Council. *Sustaining global surveillance and response to emerging zoonotic diseases.* The National Academics Press; 2009.
65. Gonsalves G, Staley P. Panic, paranoia and public health-the AIDS epidemic's lessons for Ebola. *N Engl J Med* 2014;**371**(25):2348–9. http://dx.doi.org/10.1056/NEJMp1413425. Epub 05.11.14.
66. Centers for Disease Control and Prevention (Morbidity and Mortality Weekly Report). *Support services for survivors of Ebola virus disease-Sierra Leone.* 2014. Available from: http://www.cdc.gov/mmwr/preview/mmwrhtml/mm6350a6.htm.

Chapter 11

Treatment of Ebola Virus Disease: Therapeutic Agents

Chapter Outline

> *"We are in an unusual situation in this outbreak. We have a disease with a high fatality rate without any proven treatment or vaccine,"* says Dr Marie-Paule Kieny, assistant director-general at the World Health Organization (WHO). *"We need to ask the medical ethicists to give us guidance on what the responsible thing to do is."*[1]

WHO published a report on the ethics of using unregistered interventions to treat Ebola virus disease, where they concluded "In the particular context of the current Ebola outbreak in West Africa, it is ethically acceptable to offer unproven interventions that have shown promising results in the laboratory and in the animal models but have not yet been evaluated for safety and efficacy in humans as potential treatment or prevention."[1]

THE EVOLUTION OF ANTIVIRAL AGENTS

Viruses, unlike other microbes, are not made of cells. They are actually genetic material wrapped in a case of proteins. They have no function as long as they are outside of cells. Viruses enter cells when their protein case comes into contact with the outer surface of a human cell, and only then, they can take over cell function. The virus, once inside a cell, hijacks the machinery of a cell to replicate, and eventually replicates so many copies that the cells burst and release many copies of that virus only to wreak havoc in adjacent cells.

William Herman Prusoff (June 25, 1920–April 3, 2011), a pharmacologist, was among the first to develop an effective antiviral agent, approved in the

1960s by Food and Drug Administration (FDA).[2] Dr Prusoff spent most of his career in studying thymidine, a building block of deoxyribonucleic acid (DNA).[3] In the latter part of the 1950s, for the first time a clinical antiviral drug was shown as having selective antiviral activity when used properly. William Prusoff developed the first thymidine look-alike, 5-iododeoxyuridine, and Professor Herbert E. Kaufman showed the compound to be an effective treatment for herpes of the eye by disrupting the virus' ability to reproduce in topical form. It worked by acting as a decoy genetic material. This decoy material does not allow the cells machinery to work properly, hence rendering the virus ineffective and serving as a treatment. So if a virus was replicating in a human, the drug would stop it in its tracks. This discovery was a big scientific breakthrough as it opened doors for a new era of antiviral therapy.

A drug by the name of acyclovir was the next big antiviral agent in the game. The drug worked by stopping the virus from using its mechanisms to replicate its genome. It was derived from a sponge that resides in the Caribbean seas, *Cryptotethya crypta*. Howard Schaffer and Robert Vince codiscovered the compound, which proved to have promising antiviral activity, and patented by Schaffer in 1979. A researcher by the name of Richard Whitely would be the first to use the drug in patients. Acyclovir is still used today, especially for infection by Herpesviridae.[4]

Therapeutic development was a challenge, especially for specific agents targeting Flivovirus group, Ebola, and Marburg viruses even as early as in the 1990s. On the other hand, other viral hemorrhagic fevers like Lassa fever and Crimean–Congo hemorrhagic fever responded to a drug called ribavirin. Ribavirin works by acting as a decoy viral genome and, when incorporated into the genetic material, does not allow for further manufacturing of viral copies. Since ribavirin was effective against other hemorrhagic fevers, it made sense to run an experiment to see if it had potential use as a therapeutic against Ebola virus. Unfortunately, looking at its activity in laboratory viral cultures, it proved to be useless against the Ebola virus.[5]

In August of 2011, the Massachusetts Institute of Technology (MIT) discovered a drug that would only attach virally infected cells. They would call this drug Double-stranded RNA Activated Caspase Oligomerizer (DRACO). The drug would program only virally infected cells to die. It proved to work against all 15 viruses it was tested against, including viruses like polio virus and dengue fever.[6,7]

ANTIVIRAL AGENTS FOR EBOLA VIRUS

Now that the Ebola virus disease has spread across West Africa and to different continents, therapeutic intervention is desperately needed. Since Ebola virus disease is extremely rare, its rarity has not allowed for proper testing of effective treatment. Now that were in the largest outbreak to date and the entire globe is at risk, rather than just a remote village in Africa, the race to develop drugs against Ebola virus is on.

The most notable antiviral agent to date is the drug brincidofovir discovered by Dr Hostetler, a professor of medicine at the University of California, San Diego School of Medicine. Chimerix, founded by Hostetler in 2002,[8] announced that brincidofovir would be one of two investigational agents clinically tested in West Africa in patients with Ebola virus disease. This drug had actually been developed for the treatment of another virus, cytomegalovirus, which is well known for its infection in patients with weak immune systems. It is currently in Phase 3 trial for the treatment of adenovirus and cytomegalovirus after proving to have a good safety profile and works by stopping the replication process of viruses, specifically DNA viruses. Paradoxically, it has been shown to be effective against Ebola virus, which is an RNA virus. Essentially, the reason it interferes with Ebola virus survival is unknown. Brincidofovir has also been stockpiled by the United States as potential use as a biodefense agent, especially against smallpox.

How exactly did this drug come into the Ebola virus disease scene? The Center for Disease Control and Prevention (CDC) and the National Institutes of Health asked a drug manufacturer, Chimerix, to give them large quantities of antiviral drugs to test in virus cultures in a laboratory setting. Incidentally, brincidofovir was a potent inhibitor of Ebola virus in culture.[8] In October 2014, the FDA provided emergency authorization for use of this agent in patients with Ebola virus disease.[9] Although there are many other antiviral agents being developed for the Ebola virus disease, brincidofovir had moved ahead of the pack because of its established reputation. The drug has already been used in hundreds of human subjects including children, and health professionals know its properties. Brincidofovir has been used in two patients thus far, Thomas Duncan and Ashoka Mukpo and, although no evidence reports its efficacy in humans, the drug company is ready to start trials for patients. Unfortunately, Duncan received the therapeutic agent when he was already critically ill and passed away. Ashoka Mukpo fortunately survived.

Another drug developed by Tekmira (Tekmira Pharmaceuticals Corporation) called TKM-Ebola, is already being tested in a Phase 1 trial.[1] However, the trial was put on a partial hold because some subjects had a flu-like response to the medication. The medication can still be used with expanded access in patients with Ebola virus disease.

Other investigators have proposed that since Ebola virus disease causes the victims to bleed, why not just stop the bleeding? With that in mind, a publication in 2012 described using a drug that was derived from nematodes. This drug works by acting on the mechanisms in blood that control the balance of its inherent clotting properties. Since Ebola virus disease patients are likely to bleed, it made sense to use a drug that gives the blood a tendency to clot in order to counterbalance the bleeding. The drug resulted in survival in three of nine rhesus monkeys.[10] In the same publication, the development of two new therapeutic agents AVI-7537 and AVI-7288 were described. Both of these agents are actually small strands of genetic material developed by Sarepta Therapeutics Inc. These drugs worked by stopping the virus from making important products essential to the virus's inherent function, therefore, rendering it nonviable. The drug had a promising safety profile in one study, but further studies could not be done, possibly due to a lack of funding.[10]

VACCINES

Vaccines, in general, are used to prevent infection. Vaccines are theoretically effective before an individual is infected with an agent, in this case, Ebola virus. Now how are vaccines made? And how do they work? The answer lies in the structure of the Ebola virus.

The Ebola virus, as mentioned above, is made of genetic material wrapped in a protein coat. The components, when broken down, are essentially harmless to humans, it is the entirety of the intact virus inside that raises hell during an infection. The body recognizes the various components of the virus and produces a reaction to those components. With those facts in mind, why not have some drug companies obtain samples of Ebola virus, break the virus down into harmless components, and inject the "component" particles into humans, so that we can mount a reaction to them?

These are the basic questions asked when a vaccine needs to be made. Once the body is exposed to Ebola virus particles, it will make antibodies that react to them, as if it is actually fighting the virus. Antibodies need to be "specific" when attacking an invader. The body has to manufacture a different set of them for each unique infection. Once the body has fought off the infection, it "learns or remembers" the structure of the invader it has fought off, so that, in the event it invades again, there is stored ammunition against it. A vaccine is a mock of the virus that does not cause infection, but do allow the body to manufacture the necessary ammunition in the event of a true Ebola infection ensues.

The U.S. National Institute of Allergy and Infectious Diseases (NIAID) and GlaxoSmithKline (GSK) have teamed up to create vaccines that are currently being tested.[11] The vaccines have completed their first phase of testing on 20 healthy individuals in the United States to determine their safety and immune response. This vaccine was actually made from a virus that causes colds in monkeys. The cold-causing adenovirus was engineered to express components of Ebola virus, hence "mimicking" Ebola virus. A similar vaccine was also made for another strain of Ebola virus and is currently being tested in Mali, the United Kingdom, and Switzerland. The investigators also tested different doses in the individuals and found that there are more antibodies made when a higher dose is given compared to a lower one.

CHALLENGES TO VACCINE DEVELOPMENT: LESSONS FROM MALARIA AND HIV INFECTION

Challenges in developing vaccines for Ebola virus infection can be explained by lessons learned from the past. The challenges originate from many areas;

perhaps the most challenging are the properties of the intruder itself. A good example of this is malaria. The malaria parasite, which is transmitted by mosquitoes to humans, infects over a couple million people and kills just under a million people per year. Scientists have been searching for a vaccine for years, but the solution has evaded them. The real issue lies in a mechanism the parasite possesses called antigenic variation. Antigenic variation means that the parasite is able to change the way it looks to the immune system. Anytime the immune system attempts to mount a response to the parasite, the parasite will change itself like a chameleon (*Chamaeleonidae*). The lizard has the ability to change colors according to the surrounding habitat and go undetected. Similarly, the malaria parasite can go totally undetected, therefore, continuing to thrive and devour its host. Since vaccines are basically decoy intruder molecules, developing a vaccine for something that has no true constant recognizable identity to the immune system remains a challenge.[12] Because of their complex life cycle, complicated genome, and the poorly understood response of the human immune system to malaria, the fight to develop a vaccine continues.[13]

Another virus that we just cannot seem to develop a vaccine for is human immunodeficiency virus (HIV), despite it being around for over 30 years. Even with the NIAID's worldwide effort to create a vaccine, there have been no effective vaccine. HIV is a bit different from most viruses, in the sense that the human body can never rid itself of HIV. Other viruses, like the flu and Ebola viruses, can be cleared by the human system, and often patients develop immunity to the viruses. The antibodies that the body produces against HIV are not effective in fighting the virus and the other arms of the immune system that do not use antibodies to fight infection are attacked by the HIV virus itself. Not only is the immune system not effective against the virus, but HIV attacks the immune system itself, making the body more prone to other infections, which is often the cause of death in HIV-positive patients.[14] Even if a vaccine were to be developed, the complex life cycle of HIV poses a challenge because HIV is a "retrovirus." Retroviridae have the unique ability of being able to incorporate their genome into the genome of the host. Meaning that HIV basically can hide within the genome of human cells, making it extremely difficult to target.

Now that the potential challenges have been explained, let us discuss the Ebola virus vaccine. The first and foremost issue is the rarity and combined lethality of the virus. The outbreaks tend to be small and in remote areas of Africa, usually self-limiting after a hundred cases or so. Subsequently, information, the priority of research and therapeutic development has not entered the priority of big player pharmaceuticals, at least until this outbreak. Another deterrent is the daunting process of development. Once someone has come up with a potential candidate vaccine, the rigorous process of testing and approval ensues. Some drug trials can take up to 10 years, if not longer. The safety and efficacy has to be thoroughly tested for the vaccine to be approved. Although the process is long and arduous, it is in the best interest of vaccine recipients.

But what happens when the situation demands a vaccine for a deadly hemorrhagic virus spreading with no regard for life and has essentially become a global threat? Even with this current scenario, "fast-track" development projections for current vaccines are going to take a year or so. And although this is exceptionally fast in comparison to other therapeutic approvals, is it fast enough to stop or slow the Ebola virus disease? Only time will tell.

Although the current vaccine has proved to induce development of antibodies in humans, we still do not know that these antibodies are going to be effective. Is Ebola virus going to be like a chameleon and change itself to evade the immune response? We already know that the Ebola virus can target the immune system, rendering it ineffective. There are multiple strains of Ebola virus. Can a single vaccine protect against all strains? Can a vaccine be developed such as the quadrivalent flu vaccine which protects against two influenza A viruses and two influenza B viruses. What if the developed vaccines prove to not have a true efficacy after being deployed on a large scale? We may have to start from ground zero yet again.

ANTIBODIES FROM EBOLA VIRUS DISEASE SURVIVORS

Antibodies made by survivors of Ebola virus disease continue to circulate within the blood for an undetermined amount of time. These antibodies can be extracted from blood and given to patients who actively suffer from Ebola virus disease. The extracted product is called convalescent serum. In order to prepare this serum, blood must first be extracted from Ebola virus disease survivors. Once blood is collected, it must be spun in a centrifuge to separate the components. The heavier elements, including red and white blood cells and platelets, will naturally separate from the lighter component, plasma. It is the plasma that contains the convalescent serum. The plasma is then transfused into a suffering patient. In Kikwit, Democratic Republic of the Congo, between June 6 and 22, 1995, eight patients were transfused with blood donated by five convalescent patients who met the case definition used in Kikwit for Ebola virus disease. The donated blood was expected to contain antibodies against Ebola virus. All the transfusion recipients were tested positive for Ebola virus antigens just before the transfusion. The transfused patients had clinical symptoms similar to other Ebola virus disease patients during the epidemic. All were severely weak, four had hemorrhagic manifestations and two became comatose as their disease progressed. Compared to overall very high case fatality rate (80%) for Ebola virus disease epidemic in Kikwit and the fatality rates observed in other Ebola virus disease epidemics, only one of the patient who received the survivor antibodies died.[15] The cause for this low fatality rate remained unexplained, and results are confounded by better care which was given to transfused patients as compared to those treated in the initial phase of the epidemic, which could have contributed to the higher survival rate.[15]

ANTIBODIES MADE FROM OTHER ORGANISMS

Instead of waiting for a survivor to appear with antibodies, why not just infect other living organisms with Ebola and extract their antibodies? The first drug used experimentally against Ebola virus disease was an agent called ZMapp developed in the tobacco plant *Nicotiana benthamiana* in the bioproduction process known as "pharming."[16] ZMapp is a drug cocktail of antibodies made against Ebola Virus. Leaf Biopharmaceutical (LeafBio, Inc.), a San Diego-based arm of Mapp Biopharmaceutical is developing the composite drug.

An experiment published in *Nature* indicated that, if ZMapp treatment is initiated up to 5 days post-Ebola virus infection, the antibody cocktail is able to save 100% of rhesus macaques. In many animals, high fever, viremia, and abnormalities in blood count and blood chemistry were evident before ZMapp intervention. Elevated liver enzymes, mucosal hemorrhages, and generalized small bleeds, indicating the advanced disease could be reversed, leading to full recovery. ZMapp is cross-reactive with the Guinean variant of Ebola virus as demonstrated by enzyme-linked immunosorbent assay and neutralizing antibody assays, which is a test used to detect microbes. Further development of ZMapp for clinical use is based on its surpassing value over any other therapeutics described so far.[17]

Despite ZMapp promising to be a useful therapeutic agent as demonstrated in the experiment with rhesus monkeys, an enormous supply and demand issue has emerged. The reason for this shortage lies in the development process of ZMapp. To produce this drug, genes encoding the antibodies are inserted into viral vehicles, which are then used to deliver the genome to tobacco plants, and, in turn, this produces the antibodies from the delivered genes. The antibodies are than extracted and purified for development. This entire cycle takes a few months to complete. The amount of ZMapp needed for this current outbreak is severely outpaced by the spread of Ebola virus disease. Even once there is enough ZMapp made, its true effectiveness in humans has yet to be determined, although it has proved valuable in rhesus monkeys. It would have to go through a similar vigorous testing process like other vaccines. We can only hope that popular approval and large-scale production will catch up to the wrath of the current Ebola virus disease epidemic.

At this current point in time, there is no proven vaccine or treatment for Ebola virus disease. The current treatment is purely a supportive one. The therapeutic agents mentioned above have been used in several patients in the United States, and a majority of those patients have survived. More studies are required to know the toxicity profile for current experimental therapeutics.

REFERENCES

1. Kroll D. *FDA moves on Tekmira's Ebola drug while Sarepta's sits unused Forbes*. August 7, 2014. Available from: http://www.forbes.com/sites/davidkroll/2014/08/07/fda-moves-on-tekmiras-ebola-drug-while-sareptas-sits-unused/.
2. Encyclopedia tf. Prusoff W. 2014. Available from: http://en.wikipedia.org/wiki/William_Prusoff.
3. Cheng Y-C. William H. Prusoff (1920–2011): father of antiviral chemotherapy. *PLoS Biol* November 8, 2011. Available from: http://www.plosbiology.org/article/info%3Adoi%2F10.1371%2Fjournal.pbio.1001190.
4. De Clercq E, Field HJ. Antiviral prodrugs – the development of successful prodrug strategies for antiviral chemotherapy. *Br J Pharmacol* January 29, 2009. Pubmed Central PMCID: 16284630. English.
5. Huggins JW. Prospects for treatment of viral hemorrhagic fevers with ribavirin, a broad-spectrum antiviral drug. *Rev Infect Dis* May–June 1989;**11**(Suppl. 4):S750–61. PubMed PMID: 2546248. Epub 1989/05/01. eng.
6. Rider TH, Zook CE, Boettcher TL, Wick ST, Pancoast JS, Zusman BD. Broad-spectrum antiviral therapeutics. *PLoS One* July 27, 2011. Available from: http://www.plosone.org/article/info%3Adoi%2F10.1371%2Fjournal.pone.0022572.
7. Trafton A. *New drug could cure nearly any viral infection*. MIT News; August 10, 2011. Available from: http://newsoffice.mit.edu/2011/antiviral-0810.
8. Forbes. *Chimerix's brincidofovir given to Dallas, Nebraska Ebola patients*. 2014 [cited 16.12.2014]. Available from: http://www.forbes.com/sites/davidkroll/2014/10/07/chimerixs-brincidofovir-given-to-dallas-nebraska-ebola-patients/.
9. Inc C. Chimerix announces emergency investigational new drug applications for brincidofovir authorized by FDA for patients with ebola virus disease [cited 16.12.2014]. Available from: http://ir.chimerix.com/releasedetail.cfm?releaseid=874647.
10. Iversen PL, Warren TK, Wells JB, Garza NL, Mourich DV, Welch LS, et al. *Discovery and early development of AVI-7537 and AVI-7288 for the treatment of Ebola virus and Marburg virus infections*. 2012. Available from: http://www.mdpi.com/1999-4915/4/11/2806/htm#sthash.8AAfRCYH.dpufhttp://www.mdpi.com/1999-4915/4/11/2806/htm.
11. NIAID. *NIH to launch human safety study of Ebola vaccine candidate*. 2014 [cited 16.12.2014]. Available from: http://www.forbes.com/sites/davidkroll/2014/08/07/fda-moves-on-tekmiras-ebola-drug-while-sareptas-sits-unused/.
12. Giles C. *Why don't we have a malaria vaccine?* September 27, 2005. Available from: http://malaria.wellcome.ac.uk/doc_wtx033040.html.
13. Prevention CfDca. *Malaria vaccines*. 2014. Available from: http://www.cdc.gov/malaria/malaria_worldwide/reduction/vaccine.html.
14. AIDS.gov. *Vaccines*. 2014. Available from: http://aids.gov/hiv-aids-basics/prevention/prevention-research/vaccines/.
15. Mile 2 Herald. How African doctors defeated Ebola in the Democratic Republic of Congo… cases of 8 patients from August 2, 2014. Available from: http://mile2herald.wordpress.com/2014/08/02/how-african-doctors-defeated-ebola-in-the-democratic-republic-of-congo-cases-of-8-patients-from-a-medical-journal-1/.
16. The Free Encyclopedia. *ZMapp*. 2014. Available from: http://en.wikipedia.org/wiki/ZMapp.
17. Vaccines and Global Health: Ethics and policy. *Ebola outbreak shuts down malaria-control efforts; Reversion of advanced Ebola virus disease in nonhuman primates with ZMapp*. October 2, 2014. Available from: http://centerforvaccineethicsandpolicy.net/2014/10/06/ebola-outbreak-shuts-down-malaria-control-efforts-reversion-of-advanced-ebola-virus-disease-in-nonhuman-primates-with-zmapp/.

Chapter 12

Ebola Virus: Natural Reservoirs

Chapter Outline

The history of the interaction between animals and humans can be traced to Cain, a farmer, and Abel, a shepherd. One day, each offered a sacrifice to God; Cain brought the fruit of the ground, and Abel brought a perfect lamb. God accepted Abel's offering, but not Cain's. Cain's produce was not worthy, according to God's command, and was the work of his own hands. Cain's fury led to murdering of his brother Abel. Adam and his help mate were sitting, weeping, and lamenting over Abel, and they did not know what to do with Abel, for they were not acquainted with burial. A raven, one of whose raven companions had died, came and dug in the earth and buried the other raven before his eyes. Adam said, "I shall do as this raven." Immediately, he took Abel's corpse and dug in the earth and buried it (Genesis 4:1–17, King James version). Thus, the bond between animal and humans was formally sealed by myth or belief and evolved into ideas about good and evil for centuries to follow.

Without domesticated and wild animals, the survival of mankind would not be possible. But animals have been a harbinger of death for humans outside of isolated wrath of an animal turning to survival instincts. Many of the diseases that have plagued humanity are caused by microorganisms that originated from other species of animals. When humans began to domesticate animals for agricultural purposes, some of those microorganisms mutated into forms that could cause disease in humans. Diseases transmitted from animals are called zoonotic diseases and include hantavirus pulmonary syndrome, bubonic plague, tularemia, West Nile virus, western equine encephalitis, Colorado tick fever, and rabies. These are carried by various animals and insects (rodents, mosquitoes, or ticks) and then transmitted to humans. Measles, smallpox, influenza, diphtheria, and tuberculosis probably came to humans from other animals and are now transmitted directly from human to human. Other zoonotic diseases are still transmitted by animals. Yellow fever and malaria, as example, are transmitted by mosquitoes. Bubonic plague is carried by rats (and is also highly contagious among humans), and rabies can be carried and transmitted by any mammal.

ZOONOTIC CONTRIBUTION TO THE NEW WORLD EPIDEMICS

By the fourteenth century, Europeans, Asians, and Africans had extensively domesticated animals and acquired several endemic diseases. As a result, Europeans, Asians, and Africans had developed a natural resistance to smallpox, measles, typhoid, and other diseases endemic to the Old World. Many of the zoonotic diseases that were well established in the Old World were absent from the Americas before the arrival of Christopher Columbus in 1492. American Indians, who lived on a continent where the smallpox virus and other Old-World microbes did not exist, had no natural resistance to them. The catastrophic epidemics that accompanied the European conquest and the arrival of domesticated animals in the New World decimated the indigenous population of the Americas.

Five hundred years have passed and yet zoonotic diseases continue to evolve and emerge. Natural reservoir or *nidus,* the Latin word for "nest," is defined as a sustained host of a pathogen or infectious disease. It is not just the ability to carry the infection but to spread the infection without showing any signs of the disease in the carrier or host. Isolating natural reservoirs of different infectious diseases is one of the most effective ways to control the spread of disease and prevent future outbreaks. Many scientists and hypothesized that wild animals are the natural reservoir of Ebola virus though ambiguity continues to exist regarding whether one animal or species is the true reservoir of this virus. If studied further, locating the animal or species may hold the key to controlling future outbreaks in West and Equatorial African countries.

LINKING EBOLA VIRUS DISEASE TO ANIMAL HOSTS

The first step in implicating animals in the spread of any disease such as the Ebola virus infection is that the animal should have evidence of the virus' presence. When the outbreak in Gabon occurred in 2001–2003, a large team of researchers comprised of members of World Health Organization, Viral Hemorrhagic Fever Committee, and many others developed a method for detecting infection in animals. Since the disease was also found to be fatal in local ape population, including chimpanzees and gorillas, any animal carcass that was found, whether ape or not, was sampled by the research teams. When a carcass was found in good condition, specimens of liver, spleen, muscle, and skin were taken. These samples were set in cryopreservation ($-197\,°C$) for later testing by multiple techniques including polymerase chain reaction (PCR), antigen detection, and immunohistochemical staining.[1]

A majority of live animals tested were bats because they were implicated as being the natural reservoirs of the Ebola virus. Live bats were captured in nets and later taken to research laboratories where they were euthanized.

After, a detailed examination of the bat was done from weight and size to sex and species. Blood samples were taken by cardiac puncture and plasma was also obtained by centrifugation. Following necropsy, researchers obtained spleen and liver samples to be tested for the virus.[2] The results are described in subsequent sections.

EBOLA VIRUS DISEASE AND NONHUMAN PRIMATES: CARRIERS OR VICTIMS?

Speculations started as the Ebola virus disease emerged in areas of Sudan and Zaire (now Democratic Republic of Congo, DRC) and the first human Ebola virus disease cases had been exposed to suggested reservoirs of monkey and bats. In the Sudan outbreak of 1976, the first three patients who presented symptoms worked in a cotton factory which were recorded to have been infested with rodents (*Rattus rattus*) and the roof spaces of the factory was infested with a large population of insectivorous bats, Trevor's free-tailed bat (*Mops trevori*).[3] Similarly, the first reported case in Zaire in 1976 was a man who, during his travels, had bought fresh antelope and monkey meat.[4]

Over the years, there have been other outbreaks, many of which have been linked to hunters who returned with the disease and caused further outbreaks. Those hunters were suspected to have been exposed to chimpanzees or gorillas that had also contracted the virus.[5] Though chimpanzees and gorillas have been implicated as reservoir host of Ebola virus, scientists struggle for an explanation for the infected ape species succumbing to the disease and dying rapidly prior to establishing contact with human and thus do not meet the definition of a natural reservoir.[5,6] Skeptics argue that contact between nonhuman primates and humans has been integrated into the daily life of Africa for centuries, and ask why now and why here for a population that should have immunity?

The Guardian, a British national daily newspaper, published interesting article in the early 1970s describing the suspected Marburg virus disease as the cause of infection to the staff of Ministry of Defense microbiological research team in Porton Down, Wiltshire. The article was titled "'Monkey Virus' Threat to Porton workers." After the first outbreak of Marburg virus in 1967 in Germany, the name "green monkey disease" was given to the Marburg fever. "Green monkey" because it was thought to have been transmitted by vervet monkeys (*Chlorocebus pygerythrus*; Figure 12.1) and their characteristic golden–green fur. There was some confusion when another article was published titled "Monkey Disease," relating the disease to the Ebola virus disease outbreaks in Zaire and Sudan in 1976. The overlap of green monkey disease in both Ebola and Marburg outbreaks suggests that this could also be a potential reservoir of both viruses.

Green monkey disease was also studied in 1969 when the outbreak of 1967 was researched. Interestingly, this research attributed the green monkey disease to a subspecies of the vervet monkey called the *Cercopithecus aethiops*. These monkeys also have the distinct golden-green colored fur. These were used in the production of vaccines for diseases such as poliomyelitis and rubella at the time.[7]

FIGURE 12.1   Green monkeys (*Chlorocebus pygerythrus*).

It has also been suggested that, though green monkeys (*C. aethiops*) do not directly cause human immunodeficiency virus, they have been shown to transmit the simian immunodeficiency virus to rhesus macaques, which induced an acquired immunodeficiency virus-like syndrome in them.[8]

During the outbreaks in Gabon between 1994 and 2003, there was a report of about 397 dead animal carcasses, including gorillas, chimpanzees, mandrills, and bushpigs. During the same time, another 14 carcasses tested positive for Ebola virus in the areas of northwestern Republic of Congo and northeastern Gabon.[9] Was this an example of epidemic among nonhuman primates, with the virus being transmitted between species? The large number of deaths again exonerated nonhuman primates from being a reservoir. It has been proposed that studying ape behavior and feeding habits may also help understand interspecies spread of the Ebola virus, due to the fact that gorillas and chimpanzees may share a common habitat.[10]

The 14 carcasses that tested positive for Ebola virus in the areas of northwestern Republic of Congo and northeastern Gabon led to an organized collaboration between 2001 and 2003. With the help of the Gabonese and Republic of Congo governments, an Animal Mortality Monitoring Network (AMMN) was set up to recruit local hunters and other wildlife organizations to gather animal carcasses of suspected Ebola virus infected cases. During this study, 98 carcasses were collected, and the global positioning system locations of carcasses were recorded. All tissues and blood samples were tested for Ebola virus using multiple techniques including PCR, antigen detection, DNA amplification, and immunohistochemical staining. In the end, 10 gorilla species (*Gorilla gorilla*; Figure 12.2), 3 chimpanzees (*Pan troglodytes*; Figure 12.3), and 1 duiker tested positive for Ebola virus. The positive strain of the virus was the Zaire strain of Ebola virus.[1] Further studies of the resident gorilla population actually showed a much larger scale of infection caused by the Ebola virus, where an estimated between 5500 and minimum 3500 gorillas died secondary to the disease in the outbreaks that began in late 2001–2002, assuming that the ape density in Ebola virus-stricken areas was about 4.4 gorillas/km^2.[11]

FIGURE 12.2   Gorilla (*Gorilla gorilla*).

FIGURE 12.3   Chimpanzee (*Pan troglodytes*).

There was one type of Ebola virus known as Ebola-Reston identified in Virginia, United States, in 1989 among cynomolgus monkeys (*Macaca fascicularis*) imported from the Philippines. Out of the 10 animals tested for the presence of virus, 5 demonstrated positive results. A later shipment of monkeys was also evaluated for high mortality and when serum and tissues samples (mainly from liver) were tested, Ebola virus antigen was detected in seven of them. Techniques such as electron microscopy to visualize morphology of the virus as well as immunohistochemistry for viral antigen were used. No human who was in contact with the monkey ever developed the disease.[12]

EBOLA VIRUS DISEASE AND BATS

Although considered pests themselves with a general public fear of spreading many diseases, bats play a very integral role in our ecosystem. They have been shown to cause arthropod suppression, which helps to control agricultural pests, to consumption of mosquitoes, which is a major cause of malaria in developing countries. Fruit-eating bats also play an important role in the pollination and seed dispersal of many plants some of which may be endangered by deforestation. Guano, or bat droppings, plays a key role in agriculture where it is mined from bat caves and used as a fertilizer due to the presence of high concentrations of nitrogen and phosphorus shown to be beneficial to plants.[13]

The suspicion that bats are natural reservoirs of Ebola virus is supported by studies performed from the bat family of *Filoviridae*. Several bats tested positive for virus-specific antibodies in sera, and the virus was detected in bat tissue. The tests were done using qualitative real-time polymerase chain reaction (Q-RT-PCR) to detect Marburg virus nucleic acid and other tests such as IgG enzyme-linked immunosorbent assay (ELISA) and immunohistochemistry on formalin-fixed kidneys. Histopathological studies were also done on liver and spleen, which displayed the presence of Marburg virus in hepatocytes. These positive results were gathered from Egyptian fruit bats (*Rousettus aegypticus*) residing in Kitaka Caves in Uganda where miners were diagnosed with Marburg hemorrhagic fever in 2007.[14]

There was also a study performed between 2001 and 2003 where Ebola virus-specific IgG immunoglobulin antibodies and Ebola virus ribonucleic acid (RNA) were obtained in about 6% of the total bats that were studied in Gabon and Democratic Republic of Congo (DRC).[15] One of the most recent studies was carried out in DRC subsequent to Ebola virus disease outbreak between June 2003 and May 2005 and then again a year later in May of 2006. The bats were captured using mist nets set up during evening when bats usually come out to feed. Trapped bats were collected in the morning and transported to field labs where they were euthanized and studied. During this study, bats were also collected from Gabon in February 2005, in October 2006, and in March 2008 and tested. The researchers identified Zaire Ebola virus strain-specific IgG immunoglobulin antibodies, confirming exposure to virus in nine different bat species. Most notable were four species namely *Hypsignathus monstrosus* (hammer-headed bat; Figure 12.4), *Epomops franqueti* (Franquet's epauletted fruit bat; Figure 12.5), *Myonycteris torquata* (little collared fruit bat), and *R. aegypticus* (Egyptian fruit bat).[2]

Both Ebola and Marburg viruses were isolated from one bat species *R. aegypticus*. The concurrent existence led to speculations that Ebola virus may be a newly mutated form of the Marburg virus. Structural similarities between the two viruses would also lend some support to the theory of the evolutionary process.

Presence of Ebola virus without contracting the disease in bats may be related to their ability to exhibit torpor and hibernate. During this period, it might be possible for bats to continue to shed the virus and infect other bats in the same group without enough metabolic activity to face consequences of the virus pathogenicity. Bats are one of the earliest mammalian species that have continued to evolve and their immune systems may have incorporated evolutionary changes that are lacking in other mammalian species, causing bats to be the reservoirs and not manifest the disease.[16]

FIGURE 12.4   Hammer-headed bat (*Hypsignathus monstrosus*).

FIGURE 12.5   Franquet's epauletted fruit bat (*Epomops franqueti*).

However, just because bats had evidence of Ebola virus exposure does not mean that they passed this virus to humans. To further evaluate the transmission process, data were collected from villages, hunters, and surrounding areas of previous outbreaks and ongoing outbreaks. During the outbreak in DRC in 2007, a detailed study was done outlining the potential mode of transmission from bats to human. It was speculated that shortly after DRC gained independence, most residents of remote bush villages were moved closer to main roads for better access to health care, education, and other public services. This reorganization of residences could have potentially been a major cause of further spread of the disease. Villages in DRC are set up in a chain fashion in which there is a linkage between two villages, one in the bush and the other near main roads and close to cities. The bush village is where hunters bring bushmeat to be prepared and then the meat shipped to secondary village which is more accessible for transport.[17] The bat meat is enjoyed by all people ages and genders, including pregnant women.[17] The bats that were hunted included those previously mentioned to have been positive for Ebola virus. Such a route may allow expedient shipment of bushmeat, including bat meat, but does not explain the survival and preservation of virulence through meat processing and shipment and cooking processes. Even the hunters who use high-powered rifles for their hunting purposes have limited contact with the animals themselves. Therefore, for bats to be responsible for spread of Ebola virus, the virus must be transported by live bats.

There is an annual migration of bats into the areas where the outbreak occurred and during this particular outbreak according to the villager there were unusually large quantities of bats.[14] Bats are the only mammal species capable of flying across large geographical areas and migratory paths. In addition, when these bats stop to rest, they cause food particles and excreta to fall on forest floor where other susceptible animals and humans are exposed and eventually exhibit the disease. The relative long life span of bat species may also contribute to virus replication and infectivity. R0, which is the number of newly infected hosts that one infectious host will produce during its period of infectiousness, is likely to be high for bats.[16]

CAMPAIGN AGAINST AFRICAN BUSHMEAT

During the current epidemic of Ebola virus disease, government and health authorities have tried to restrict or even prohibit the acquisition and consumption of animal meat in many countries including many West African countries due to the new scare of Ebola virus. Efforts have ranged from forced closing of sales of bushmeat in local markets in Guinea, Sierra Leone, and Cote d'Ivoire to raids in New York City and Minneapolis in shops that may have imported bushmeat. Bushmeat is defined by the Center for Disease Control and Prevention (CDC) as any meat that potentially comes from wild animals captured in developing nations such as Africa. It is considered illegal to bring any kind of bushmeat to the United States. If such meat is found, it is destroyed and suppliers face a $250,000 fine for bringing bushmeat to the United States.[18]

Bushmeat is considered delicacy by majority of the African population. New York City is home to nearly 77,000 West Africans and is considered the epicenter of bushmeat trade. According to the U.S. Department of Agriculture rules, no African country is allowed to import *any* type of meat product. Despite these laws, there has been confiscation of over 69,000 different bushmeat items, and an estimated 15,000 pounds is smuggled in every month. Bushmeat can also range from bat to monkey to even lion meat.

High smuggling volumes are attributed not only to the fact that it is a delicacy, but also to the African belief that bushmeat has medicinal properties. In 2012, government officials worked with researchers at both John F. Kennedy

Airport in Queens, New York, and other smaller airports to test confiscated bushmeat and found that meat originated from many animal species including baboon, chimpanzee, mangabey, guenon, green monkey, and cane rat.[19]

The local residents of these countries remain skeptical about the possible connection between bushmeat consumption and the Ebola virus disease. There are villagers who themselves and their ancestors have consumed bushmeat for centuries and continue to avoid such diseases. The agenda of animal rights groups in rightfully preserving wildlife overflows into the current Ebola virus disease paranoia. It is undeniable that numerous victims of Ebola virus disease in West Africa never came close to a bat or nonhuman primate prior to infection. In the end, human contact supersedes any animal contact as the primary mode of transmission of Ebola virus disease.

REFERENCES

1. Rouquet P, Froment JM, Bermejo M, Kilbourn A, Karesh W, Reed P, et al. Wild animal mortality monitoring and human Ebola outbreaks, Gabon and Republic of Congo, 2001–2003. *Emerg Infect Dis* February 2005;**11**(2):283–90. PubMed PMID: 15752448. Pubmed Central PMCID: PMC3320460. Epub 2005/03/09. eng.
2. Pourrut X, Souris M, Towner JS, Rollin PE, Nichol ST, Gonzalez JP, et al. Large serological survey showing cocirculation of Ebola and Marburg viruses in Gabonese bat populations, and a high seroprevalence of both viruses in *Rousettus aegyptiacus*. *BMC Infect Dis* 2009;**9**:159. PubMed PMID: 19785757. Pubmed Central PMCID: PMC2761397. Epub 2009/09/30. eng.
3. Ebola haemorrhagic fever in Sudan, 1976. Report of a WHO/International study team. *Bull World Health Organ* 1978;**56**(2):247–70. PubMed PMID: 307455. Pubmed Central PMCID: PMC2395561. Epub 1978/01/01. eng.
4. Ebola haemorrhagic fever in Zaire, 1976. *Bull World Health Organ* 1978;**56**(2):271–93. PubMed PMID: 307456. Pubmed Central PMCID: PMC2395567. Epub 1978/01/01. eng.
5. Georges AJ, Leroy EM, Renaut AA, Benissan CT, Nabias RJ, Ngoc MT, et al. Ebola hemorrhagic fever outbreaks in Gabon, 1994–1997: epidemiologic and health control issues. *J Infect Dis* February 1999;**179**(Suppl. 1):S65–75. PubMed PMID: 9988167. Epub 1999/02/13. eng.
6. Formenty P, Boesch C, Wyers M, Steiner C, Donati F, Dind F, et al. Ebola virus outbreak among wild chimpanzees living in a rain forest of Cote d'Ivoire. *J Infect Dis* February 1999;**179**(Suppl. 1): S120–6. PubMed PMID: 9988175. Epub 1999/02/13. eng.
7. Luby JP, Sanders CV. Green monkey disease ("Marburg virus" disease): a new zoonosis. *Ann Intern Med* September 1969;**71**(3):657–60. PubMed PMID: 4980236. Epub 1969/09/01.eng.
8. Hirsch VM, Dapolito G, Johnson PR, Elkins WR, London WT, Montali RJ, et al. Induction of AIDS by simian immunodeficiency virus from an African green monkey: species-specific variation in pathogenicity correlates with the extent of in vivo replication. *J Virol* February 1995;**69**(2):955–67. PubMed PMID: 7815563. Pubmed Central PMCID: PMC188664. Epub 1995/02/01. eng.
9. Lahm SA, Kombila M, Swanepoel R, Barnes RF. Morbidity and mortality of wild animals in relation to outbreaks of Ebola haemorrhagic fever in Gabon, 1994–2003. *Trans R Soc Trop Med Hyg* January 2007;**101**(1):64–78. PubMed PMID: 17010400. Epub 2006/10/03. eng.
10. Walsh PD, Breuer T, Sanz C, Morgan D, Doran-Sheehy D. Potential for Ebola transmission between gorilla and chimpanzee social groups. *Am Nat* May 2007;**169**(5):684–9. PubMed PMID: 17427138. Epub 2007/04/12. eng.

11. Bermejo M, Rodriguez-Teijeiro JD, Illera G, Barroso A, Vila C, Walsh PD. Ebola outbreak killed 5000 gorillas. *Science (New York, NY)* December 8, 2006;**314**(5805):1564. PubMed PMID: 17158318. Epub 2006/12/13. eng.
12. Ebola virus infection in imported primates–Virginia, 1989. *MMWR Morb Mortal Wkly Rep* December 8, 1989;**38**(48):831–2, 7–8. PubMed PMID: 2511410. Epub 1989/12/08. eng.
13. Kunz TH, Braun de Torrez E, Bauer D, Lobova T, Fleming TH. Ecosystem services provided by bats. *Ann N Y Acad Sci* March 2011;**1223**:1–38. PubMed PMID: 21449963. Epub 2011/04/01. eng.
14. Towner JS, Amman BR, Sealy TK, Carroll SA, Comer JA, Kemp A, et al. Isolation of genetically diverse Marburg viruses from Egyptian fruit bats. *PLoS Pathog* July 2009;**5**(7):e1000536. PubMed PMID: 19649327. Pubmed Central PMCID: PMC2713404. Epub 2009/08/04. eng.
15. Leroy EM, Kumulungui B, Pourrut X, Rouquet P, Hassanin A, Yaba P, et al. Fruit bats as reservoirs of Ebola virus. *Nature* December 1, 2005;**438**(7068):575–6. PubMed PMID: 16319873. Epub 2005/12/02. eng.
16. Calisher CH, Childs JE, Field HE, Holmes KV, Schountz T. Bats: important reservoir hosts of emerging viruses. *Clin Microbiol Rev* July 2006;**19**(3):531–45. PubMed PMID: 16847084. Pubmed Central PMCID: PMC1539106. Epub 2006/07/19. eng.
17. Leroy EM, Epelboin A, Mondonge V, Pourrut X, Gonzalez JP, Muyembe-Tamfum JJ, et al. Human Ebola outbreak resulting from direct exposure to fruit bats in Luebo, Democratic Republic of Congo, 2007. *Vector Borne Zoonotic Dis (Larchmont, NY)* December 2009;**9**(6):723–8. PubMed PMID: 19323614. Epub 2009/03/28. eng.
18. Bringing animal products into the United States: Centers for Disease Control and Prevention; [cited 12/08/2014]. Available from: http://www.cdc.gov/animalimportation/animalproducts.html.
19. Flynn G, Scutti S. *Smuggled bushmeat is Ebola's back door to America: Newsweek.* 2014 [cited 12/08/2014]. Available from: http://www.newsweek.com/2014/08/29/smuggled-bushmeat-ebolas-back-door-america-265668.html.

Chapter 13

Economic and Political Impact of Ebola Virus Disease

Chapter Outline

We are at a dangerous moment in these three West African countries, all fragile states that have had strong economic growth in recent years after decades of wars and poor governance. It would be scandalous to let this crisis escalate further when we have the knowledge, tools, and resources to stop it. Tens of thousands of lives, the future of the region, and hard-won economic and health gains for millions hang in the balance.
Jim Yong Kim and Paul Farmer, World Bank Group President and Harvard University Professor, The Washington Post

West Africa, once home to the Ghana Empire, Mali Empire, Sosso Empire, Songhai Empire, and Akan Empire of Ashanti, is a shadow of its former glory. The fourteenth-century emperor, Mansa Musa, from West Africa was worth a staggering $400 billion, after adjusting for inflation, as calculated by celebrity net worth.[1] Musa's wealth was a result of his country's vast natural resources. The West African nation was responsible for more than half of the world's salt and gold supply. Just two generations later, his fortune was gone—wasted away by invaders and infighting.

Today, Liberia, Sierra Leone, and Guinea are low-income, neighboring countries located in sub-Saharan West Africa. Guinea, Liberia, and Sierra Leone

TABLE 13.1 Macroeconomic Indicators of the Countries Affected by Ebola Virus Infection

Affected Countries	GDP in 2013 in Current USD	GDP per Capita in Current USD	GNI[a] per Capita in 2013 in Current USD
Liberia	2 billion	454	410
Sierra Leone	5 billion	809	680
Guinea	6.2 billion	527	460

GDP, gross domestic product; USD, United States dollars; GNI, gross national income.
[a]The GNI is the total domestic and foreign output claimed by residents of a country, consisting of GDP plus factor incomes earned by foreign residents minus income earned in the domestic economy by nonresidents.
Source: World Bank.

rank among the poorest nations on earth (Table 13.1), and all three are recovering from long periods of military dictatorship (their involvement in Liberia's 1989–2003 Civil War). A period of unprecedented growth and reform was on the way prior to emergence of Ebola virus disease. Guinea held its first free election in 2010, and the International Monetary Fund had forecasted its economy would grow 4.5% that year on the back of massive iron ore deposits. Liberia had been forecasted to grow at 7% on its rising mining activity, while Sierra Leone was expected to be Africa's fastest growing economy at 13.9%, according to International Monetary Fund.

The economic indicator used to gauge the status of involved country's economy is the gross domestic product (GDP). GDP is the total dollar value of all goods and services produced over a specific time period, and percentage change can be used to gauge growth on a year-to-year basis. For example, if the GDP in Guinea is up 4.5%, this represents that the economy has grown by 4.5% over the last year. According to a report by Elvin Mayer, the GDP is calculated either by adding up what everyone earned in a year (income approach), or by adding up what everyone spent (expenditure method).[2]

Ibrahim Abdullah was a truck business owner of Lebanese descent, whose ancestors left Lebanon in the late nineteenth century during the final years of Ottoman rule.[3] He was a 5-year-old child in March 1991,[3] when an attack on a small southern village by a group of armed Sierra Leoneans, Liberians, and Burkinabes identifying themselves as the Revolutionary United Front started a 9-year-civil conflict. Like almost any family in Sierra Leon, he had lost many relatives and was forced to move to Liberia. Over the last decade, he rose through society to own a truck business comprising of a fleet of Caterpillar trucks. He had purchased three of the trucks in the early part of 2014 as part of an agreement with Western Cluster-Liberia to transport iron ore from Bomi and

Grand Cape Mount counties. ArcelorMittal, a major mining company, mined iron ore, the main ingredient in making steel and was one of his lead customers. The company's highly profitable Liberian unit produced 5 million tons of iron ore a year, at a cost of less than $30 per ton, below that of most other countries. With iron ore prices falling but still around $100 per ton that made for hefty profit margins. Therefore, Abdullah had every reason to use his lifesavings to expand the business to meet the increasing demands. When reports of Ebola virus-infected patients in Sierra Leon started to emerge, the company reaffirmed its commitment and determination to operate within the scope of the Mineral Development Agreement and ensure that infrastructures are protected as well as the well-being of the people are guaranteed.

It was a Monday morning like any other, but that morning, less than half of the drivers showed up for work. Some were sick or taking care of sick relatives, but most wanted to avoid the weak subgrade road travels through villages with Ebola virus-infected patients. With most of the available drivers willing to work longer for extra pay, Abdullah had successfully devised a plan to continue the transport routes scheduled for the week. Then he received a phone call from the representative of ArcelorMittal that the company declared *force majeure* and evacuated 645 employees from Sierra Leon.[3] Roughly 4000 other employees and contractors in Liberia stayed to secure equipment and carry out other critical activities related to logistics, engineering, and procurement.[3] Abdullah's trucking business was no longer necessary given that the much smaller amount of iron ore expected to be produced could be transported through railway tracks. Abdullah was devastated because he had overextended his company and personal resources for this venture and now had no means to recuperate from the sudden loss of revenue. He had no recourse to enforce the contracts that he had signed with the company because the legal departments of the mining giants had defaulted using the *force majeure* clause.

Force majeure is French for superior force and "is a contract provision that allows a party to suspend or terminate the performance of its obligations when certain circumstances beyond their control arise, making performance inadvisable, commercially impracticable, illegal, or impossible."[4] With the borders restricted to road traffic, little opportunities existed for Abdullah to diversify into providing logistic support for new and existing trading businesses. He would scramble over the next few months desperately trying to avoid a complete closure of his business while ArcelorMittal (MT) would go on to report a net income of $22 million or a penny per share in the third quarter of 2014, contrary to a loss of $193 million or 12 cents per share in the year-ago quarter. Revenues had gone up 2.2% year over year to $20.1 billion due to increased sales as a result of increased steel shipments and higher marketable iron ore shipments. This was the typical case for many others in Liberia, Sierra Leon, and Guinea who were crushed like ants under the feet of dancing elephants by corporate giants to whom the Ebola virus infection was merely an event (http://www-haproxy-e-c01-dev.aws.barchart.com/news/stocks/MT).

DIRECT AND INDIRECT ECONOMIC CONSEQUENCES
OF EBOLA VIRUS DISEASE EPIDEMIC

According to a report released by the World Bank, there are two distinct channels through which the outbreak could affect the economies in the Ebola virus infection-affected countries. The direct effect is probably most apparent and is based on consumption of health-care resources and removal of people either temporarily or permanently from the labor force. According to a World Health Organization (WHO) report released on September 20, 2014, in the three affected countries 2803 deaths occurred out of 5843 of total Ebola virus disease cases detected, including confirmed, probable, and suspected cases.[5] Some analysts fear that the true numbers may in fact be as large as two to four times what is reported due to underreporting issues. The major proportion of these deaths occurred in young persons, which exaggerates the workforce depletion effect. Contraction of labor resource resulted from largely workers who were diagnosed and subsequently died from the virus infection.

There was a substantial loss of productivity due to diversion of workforce by, those who are taking direct care of these patients, a phenomenon called "caregiver burden." Caregiver burden in West African countries is far greater than that seen in the United States or Europe where the social services infrastructure relieves the load borne by family members serving as caregivers. There was an unmeasured effect of people staying home due to fear of exposure to the virus. All these factors resulted in reduced productivity, lower output, and lesser household income. The reduction in workforce cannot be compensated by smaller businesses that lack the ability to provide work retribution or shed less profitable parts of the business. Thus, loss of a critical mass means a death sentence for many small businesses in these countries.

Less apparent are the indirect effects. According to a report from World Bank, in the recent history of infectious disease outbreaks such as the severe acute respiratory syndrome epidemic of 2002–2004 and the H1N1 flu epidemic of 2009, behavioral effects are believed to have been responsible for as much as 80% or 90% of the total economic impact of the epidemic.

The fear of contagion leads to a fear of association with others and ultimately decreases participation within the collective labor force. Such an effect, however, is minimized in West African countries where financial compensation relies on daily attendance, and absenteeism may result in immediate financial loss. Other behavioral effects in these countries include people who are moving from diseased zones, the decrease in labor participation mainly due to people staying home from their jobs, foreign companies scared of the disease and getting rid of their employees, and closing borders for export of various materials. Such effects also motivate private decision-makers to disrupt travel, trade, and commerce by canceling scheduling commercial flights and reducing in shipping and cargo services. Some governments are motivated to close land borders and restrict entry of citizens from afflicted countries.[6]

THE IMPACT OF EBOLA VIRUS DISEASE ON VARIOUS ECONOMIC SECTORS

Mining: The mining sector in Liberia, Sierra Leone, and Guinea is dominated heavily by iron ore mining, but also includes gold and diamond mining. The mining sector accounts for 17% of Liberia's GDP. ArcelorMittal (MT), the largest mining company in the country, decided to postpone its planned investment to expand its production capacity from 5.2 million tons of iron ore to 15 million tons, and China Union, the second largest mining company has shut down its operations since August 2014. As a result, mining sector growth forecast by the World Bank for 2014 has been revised from 4.4% growth to 1.3% contraction (Table 13.2). The mining sector of Sierra Leone, which accounts for about 16% of the country's GDP, has not been affected considerably by the Ebola virus infection outbreak, in contrast to the neighboring countries reviewed in Table 13.3. However, iron ore prices recently decreased considerably, and mining revenue is expected to decrease.[7]

Compared to Liberia and Sierra Leone, the mining sector in Guinea does not make up a major component of the economy. The initial projection of mining sector growth was −3% and the revised projection of −3.4% is only slightly worse (Table 13.4). This is also due to the fact that Guinea's major mines are far from the infected zones, and the expected contraction is not expected to deteriorate.

Agriculture: The agricultural sector has been severely affected by the Ebola virus infection epidemic in all three countries of Sierra Leone, Guinea, and Liberia. Agriculture accounts for nearly a quarter of the Liberia's GDP and employs almost half of the available workforce.[8] The major reason for the severe impact

TABLE 13.2 Liberia—Estimated Gross Domestic Product (GDP) Impact of Ebola (2014)

	Contribution to Growth Shock (%)	Initial Projection (June 2014)	Revised Projection
Real growth in GDP		5.9	2.5
Agriculture	18.0	3.5	1.3
Forestry	−0.1	2.0	2.0
Mining	27.3	4.4	−1.3
Manufacturing	4.6	9.6	5.0
Services	50.2	8.1	4.0

Source: World Bank/International Monetary Fund Staff Estimates.

TABLE 13.3 Sierra Leone—Estimated Gross Domestic Product (GDP) Impact of Ebola (2014)

	Contribution to Growth Shock (%)	Initial Projection (June 2014)	Revised Projection
Real growth in GDP		11.3	8.3
Agriculture	27.8	4.8	2.6
Industry	54.5	24.8	18.4
Mining	39.6	27.3	21.8
Services	17.7	7.7	5.7

Source: World Bank/IMF Staff Estimates.

TABLE 13.4 Guinea—Estimated Gross Domestic Product (GDP) Impact of Ebola (2014)

	Contribution to Growth Shock (%)	Initial Projection (January 2014)	Revised Projection
Real growth in GDP		4.5	2.4
Agriculture	20.3	5.7	3.3
Forestry	0.0	3.5	3.5
Mining	3.8	−3.0	−3.4
Manufacturing	2.5	6.5	5.6
Services	73.5	6.7	3.8

Source: World Bank/IMF Staff Estimates.

on both export and domestic agriculture is due to reduced workforce mobility, people migrating to safe zones, and foreign companies postponing investment due to the evacuation of key expatriates affected. As a result, the World Bank revised its growth expectations from 3.5% to 1.3% (Table 13.2). Disruptions from the outbreak during the planting season earlier this year are expected to diminish yields for the staple crops of rice and maize during the harvest season, between October and December. Due to the neglect of many small farms that produce food for domestic consumption, the World Bank expects Liberia to experience food shortages that may in turn lead to food price increases.[8]

Similarly, Sierra Leone's agricultural sector, which focuses on rice, cocoa, and palm oil, accounts for about half of the economy. According to the Ministry of Agriculture, Forestry, and Food Security, the two regions that were the epicenter of the Ebola virus infection outbreak together produced about 18% of the domestic rice output. Quarantined zones restricted worker movement and many farms were abandoned. According to the reports cited by the World Bank, rice prices jumped by 30% in the affected regions of the country.[8]

A large amount of Guinea economy is composed of agriculture and services. There was tremendous decrease in the production of cocoa and palm oil, which are the main export products of the sector and now threatens Guinea's economy.[8]

Service: The service sector accounts for about half of the Liberian economy and has been perhaps the most impacted by the Ebola virus infection outbreak. According to the World Bank data, wholesalers and retailers experienced 50–75% reduction in turnover compared to the period before the crisis. Due to reductions in business and tourist trips, there have been huge reduction in markets serving expatriates; consequently, the hotel and restaurants subsector has also been negatively affected by the crisis.

The Sierra Leone's and Guinea's service sectors face a similar crisis. Sierra Leone's tourism industry has suffered from canceled flights. Hotels are occupied at half capacity, and secondary effects have extended to the country's brewing industry. In Guinea, projected growth in the services sector has been cut in half.

OVERALL FISCAL IMPACTS

The negative effects of the Ebola virus infection outbreak on the affected countries' economies are also expected to be reflected in the fiscal balance of these countries including expected reduction in tax revenues. At the same time, governments are facing increased expenditures to meet the increased requirement for medical resources to prevent and treat Ebola virus infection. According to the World Bank, "Budget deficits of the affected countries are expected to increase by amount equal to 1.8% of gross domestic product in both Sierra Leone and Guinea and 4.7% in Liberia" (Figure 13.1). Negative effects in the agricultural sector will lead to food shortages that will cause food prices to increase. Although not anticipated at this point, food shortages can easily escalate into health consequences related to malnourishment and cast additional burdens onto a vulnerable society.

Due to contraction of the major economic sectors along with a significant decrease in exports, there will be a negative effect in GDP growth.[8] The World Bank staff has revised their 2014 GDP growth projections of the affected countries: Liberian GDP was revised from 5.9% to 2.5%, for Sierra Leone from 11.3% to 8.3%, and for Guinea from 4.5% to 2.4%[8] (see Tables 13.2–13.4).

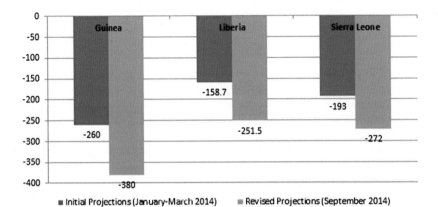

FIGURE 13.1 Estimated short-term impact on the overall fiscal balance (2014), in \$US millions. *Source: World Bank.*

FINANCIAL AID IN PROGRESS

The World Bank Group (WBG) is mobilizing a \$400 million financing package for the countries hardest hit by the Ebola virus infection crisis, including \$230 million package announced in August 2014 and an additional \$170 million package announced in September 2014.[9] Of the \$400 million committed, \$117 million is already disbursed to countries and implementing agencies, including \$58 million for Liberia, \$34 million for Sierra Leone, and \$25 million for Guinea.

WBG's Board of Executive Directors approved 105 million dollars in financial aid on September 16, and this new money provided grants for the WBG's IDA Crisis Response Window.[9] The other \$12 million in the emergency financing was reallocated in August 2014 from existing health projects in Liberia and Sierra Leone (\$6 million per country) to make some funds immediately available and take advantage of the existing implementation capacity set up for these projects. The WBG's Board of Executive Directors will consider the remaining new funding in the near future.

These funds are being used to pay for essential supplies and drugs, personal protective equipment and infection prevention control materials, health workers training, hazard pay and death benefits to Ebola virus infection health workers and volunteers, contact tracing, vehicles, data management equipment, and door-to-door public health education efforts.[9] These funds also are providing budgetary support to help the governments of Guinea, Liberia, and Sierra Leone cope with economic impact of the outbreak and are financing the scaling up of social safety net programs for people in the three countries. UN agencies—particularly UNICEF, WHO, UNOPS, United Nations Population Fund (UNFPA), and the World Food Program (WFP)—are implementing some of the activities financed by the WBG's emergency project. WBG funding is helping WFP to scale up its ongoing response to the Ebola virus infection emergency. Some examples are given below.

WBG funds have already financed a first shipment of 100 tons of essential health and hygiene supplies (worth \$1.6 million), which UNICEF procured and delivered to Liberia on August 26, 2014. Among the items in the shipment were hundreds of sets of personal protective equipment, latex gloves, thermometers, syringes, and other medical supplies to restock depleted stores at many of the country's health facilities.

On September 5, UNICEF procured and delivered 48 tons of materials and essential drugs for Ebola virus disease treatment centers (antibiotics and other essential medicines, cannula, coveralls to protect health workers, 7440 pairs of latex gloves, and body bags) to Sierra Leone (worth $850,000), also financed by the WBG.

On September 17, also with WBG financing, UNICEF airlifted a second delivery to Sierra Leone (worth $1.7 million) 100 tons of drugs and equipment, including personal protective equipment, antibiotics, intravenous fluids, and chlorine. On October 18, also with WBG financing, WFP airlifted 30 ambulances and mortuary pickup vehicles to Sierra Leone, which will help improve mobility and shorten the response time to people affected by Ebola virus disease. With WBG financing, WFP also has brought in 4000 metric tons of food to feed people in holding and treatment centers and quarantined communities in Sierra Leone, reaching 300,000 households.

On September 9, UNICEF procured and delivered 28 new, all-terrain vehicles to the Guinea Ministry of Health (worth $880,000), also with WBG financing. These vehicles will provide much-needed ground logistics support to treatment, supervision, contact monitoring, and burial teams working in Guinea. A second lot of vehicles will include ambulances for patient transfers and motorbikes for contact monitoring in remote, hard-to-reach villages. On October 15, UNICEF procured and delivered 57 tons of goods and equipment to Guinea—also with WBG financing—including 4.5 million gloves, 50 tents, medicine, and food supplements to fight Ebola virus disease.[9]

THE EFFECT OF OUTBREAK ON SURROUNDING COUNTRIES' ECONOMIES

This impact on economy is not felt in the affected countries, but this wave of economic influence is also slowing down neighboring economies, with several disruptions. Although Nigeria has been officially declared by WHO as an Ebola virus-free country, one of the reports says that "commercial businesses in Lagos indicate significant recent declines in demand, sometimes in the range of 20 to 40 percent."[10] Gambia, where the economy is largely dependent on tourism, is also experiencing a major decline in this sector. According to a Bloomberg Report, in Gambia, "hotel bookings are down by 65 percent due to a fear of the virus."[10] This is being attributed to a "geographic misfortune" of "being located within a few hundred kilometers of Senegal's border with Guinea."[10] In the unaffected countries in Africa, the stigma of the outbreak is having an impact on potential international investors and multinationals operating in these countries.[10]

In East African countries like Kenya, the geographical impact of the Ebola virus disease outbreak is also affecting major sectors like tourism and transportation.[10] However, since the recent Ebola virus disease outbreak began in December 2013, the global airline industry is perhaps the worst affected and has led to several countries imposing various rules and regulations on air travel, including strict measures like increased passenger screenings. A report by the Official Airline Guide says that, before the outbreak in 2013, 264,300 passengers traveled by air from the affected countries to either Europe or North America. However, as of October 2014, year-on-year capacity and frequency are down by 64% from the affected countries. In May 2014, there were some 427 flights from Liberia, Sierra Leone, and Guinea to any destination in the world. In October, that had fallen to some 152 flights scheduled, providing some 22,782 seats.[10]

PREGNANCIES AND CHILDBIRTHS: LONG-TERM ECONOMIC CONSEQUENCES

"A man found a dead newborn girl at a scrapyard in the ELWA Junction in Monrovia, Liberia, umbilical cord still attached. It is unknown if the baby died at the yard or if the girl was already dead. Residents said this was not the first such incident, as dumping newborns is increasing since Ebola virus disease broke out in the country."[11]

Although not uncommon in the past, dumping of newborns has greatly increased since the recent outbreak in the country. The reason is either that the pregnant patient is too afraid to visit the health-care facility due to the fear of Ebola virus disease or, if she is at home, no one will help her because they are afraid of contact with the bodily fluids. Many of the health-care clinics refused to treat pregnant women due to fear of coming into contact with blood and other bodily fluids of women, and sometimes they even do not know if the patient is Ebola virus infected or not. When nothing else works, the patients end up giving birth in the streets. According to an estimate by UNFPA, there will be more than 800,000 women who will give birth in the next 12 months, and 120,000 of these pregnant women will face complications during pregnancy and childbirth and could die if the required lifesaving emergency obstetric care is not provided. This report also states that they still need 64.5 million dollars to provide all the services required to promote health of the expectant mothers and the newborn babies in Guinea, Liberia, and Sierra Leone. Currently, they only have $3 million for this purpose.[11] If these appropriate measures are not taken, they warn that one in seven pregnant women in these West African countries could die during childbirth.[12] There may be a reduction in workforce due to newborn and infant mortality that will be felt two to four decades from now if situation does not improve.

If we go back not more than 100 years and look at the babies born during Spanish flu pandemic, we find that people who were born in the United States just after the 1918 flu pandemic or the people who were still in utero when the disease was at its peak had a higher risk of heart attack in their adulthood than those born before or after the pandemic. The researchers also found that men born in 1919 were shorter by about 0.05 inches relative to surrounding cohorts and were of lower weights as teenager with a higher percentage of various health issues.[13] In addition, children born to those women who had been infected with flu were three to seven times more likely to develop schizophrenia later in life. All of these health-related conditions caused not only an increase in economic burden of the country due to increased utilization of health-care expenses, but also led to inefficient and impaired workforce of the country. With the example of pandemic flu how the economy of the country could continue to be affected years later even after the outbreak, we hope that the Ebola virus disease outbreak is aggressively controlled; so, the duration is shortened enough to prevent unwanted health-related effects on the babies born during this period.[14]

The Ebola virus disease epidemic continues to spread rapidly in Guinea, Liberia, and Sierra Leone. If the virus continues to surge in the three worst-affected countries and spans to neighboring countries, the 2-year-regional financial impact could probably reach \$32.6 billion by the end of year 2015[9] dealing a potentially catastrophic blow to already fragile states. These economic costs can be limited if swift national and international responses succeed in preventing the spread of the virus and mitigating fear resulting from people's concerns about contagion fueling the economic impact.[12]

POLITICAL IMPACT OF THE EBOLA VIRUS DISEASE OUTBREAK

The biggest concern of the Ebola Outbreak is political, not medical.
Ashoka Mukpo [Former researcher with the Sustainable Development Institute, a Liberian Civil society advocacy group based in Monrovia]

INTRODUCTION

The Ebola virus disease outbreak has created a broad impact on the global political environment within several countries. Government's poor organization and subsequent inadequate response to the Ebola virus disease outbreak creates vulnerability to criticism and change. The crisis can promote citizens' lack of faith in their governments and aggravated social tensions.[15] Ebola virus disease outbreak clearly poses a threat to the safety and stability of Liberia, Sierra Leone, and Guinea—countries still struggling to deal with the consequences of prolonged civil wars (Liberia and Sierra Leone) and chronic political instability (Guinea). The current crisis is likely to have a noteworthy effect not only on these countries' economies, but has also affected the political and operational hierarchy within those countries.[16]

POLITICAL IMPACT OF EBOLA VIRUS DISEASE ON WEST AFRICAN COUNTRIES

Sierra Leone is one of the few countries where the political infrastructure was severely affected by Ebola virus disease outbreak. Part of the challenges to the political infrastructure were related to travel and transport limitations; restrictions on the movement of individuals; and constraints on the people's right of assembly, and community by laws which have led to a "do not touch" policy. The Sierra Leonean government has stopped spending money on services not directly related to battling Ebola virus disease and day-to-day service facilities have largely deteriorated. Open criticism of the government followed as a consequence. Sierra Leon's citizens have begun to question the president Dr Ernest Bai Koroma regarding his capacity to handle the challenges. In one such incident, a journalist was arrested after a guest condemned the President's way of handling the Ebola virus disease outbreak on his inadequate effort to institute emergency measures in the face of the epidemic.[15]

The Ministry of Health Sanitation (MOHS) has faced a lot of criticism. President Koroma has also faced disapproval for supporting Miatta Kargbo, Minister for Health and Sanitation. Kargbo's situation became increasingly weak from June 2014, after she suggested that victims of Ebola virus disease had a role in acquiring the infection and deaths could have been avoided if victims had appropriately responded. She had gone on further in criticizing health care workers for the spread of Ebola virus disease through immoral acts. In August 29, 2014, she was removed from office, with the President mentioning the need "to create a conducive environment for efficient and effective handling of the Ebola outbreak." The MOHS was quite overwhelmed by Ebola virus disease and was looking for an outside source for assistance to handle the crisis.

Fortunately, there was an improvement seen in the governance of Sierra Leone, when President Koroma handed the day-to-day running of the National Ebola Response Centre (NERC) to Major (Rtd) Alfred Paolo Conteh, Minister of Defense. Some of the initial steps taken that assured control of the government were the imposition of a 3-day, nationwide lockdown in September 2014 and sanctioning of the extension of emergency measures for another 9 months. Moreover, President Koroma's increased presence at isolation units and treatment centers and occasions celebrating survivors, improved morale. The government provided ongoing support of the development of a remote education curriculum to counter balance the schools remaining closed. The President received a cross-party support, from Sierra Leone People's Party (SLPP) leader, Brigadier (Rtd) Julius Maada Bio, and thus succeeded to create a logic of national unity.[17]

In Liberia, President Ellen Johnson Sirleaf said under the state of emergency, the government will introduce astonishing methods, if need to be the suspension of certain rights and privileges as authorized by the constitution for a 90-day period. The outbreak has exacerbated a lack of trust that many

Liberians have in their government. Because of the current uncertainties due to the Ebola virus disease epidemic, several investors, aid workers, and skilled employees started to leave the country despite continuous pleas from the Liberian finance minister for continued assistance. Elections originally scheduled for early October were continually postponed. In a nation where years of civil war have deeply undermined trust in government and its institutions, some worried a constitutional crisis would be one of Ebola virus disease's deadliest symptoms. Concerns regarding political and social unrest leading to coup and civil war still prevail.[18]

Similarly, Guinea is also facing political challenges particularly in light of upcoming elections. There are apprehensions among citizens that the delays in the organization of local elections will increase prevailing tensions between the government and the opposition parties.[16]

POLITICAL IMPACT OF EBOLA VIRUS DISEASE IN THE UNITED STATES OF AMERICA

In the United States, regardless of there being no actual risk of an Ebola virus disease epidemic, following the death of Liberian citizen Thomas Duncan (first person to contact Ebola on the United States soil) and the consequent infection of two Texan health workers who had treated Duncan, health department has been under criticism for its failure to uphold strict protocols and screening procedures for Ebola virus disease.

The United States Centers for Disease Control and Prevention (CDC) has been implicated in failure to implement preventive strategies.

On the first week of October 2014, Dr Thomas Frieden, director of the CDC issued a public statement "We will stop it in its tracks" meaning that the Ebola virus disease will not be entering the United States soil. In the ensuing months, isolated cases of Ebola virus disease were identified in the United States, leading to major embarrassment for the government. Frieden was asked to provide an assessment to the Members of Republican Party-led House committee during a 3h high-level inquiry in Washington with emphasis on his inadequate response to the Ebola virus disease outbreak, thereby even demanding his resignation. Consequently, the Obama government has faced criticism for CDC's deficient efforts. The Republicans are seizing on the issue as a chance to prove a point that President Barack Obama has not provided a real assessment of the situation by stressing Obama's recent promise that Ebola virus disease infection was "unlikely" to get to the United States. Several Republican leaders have been pressurizing the regulatory authorities to enforce new travel constraints for those countries in Africa where the outbreak began.

On October 17, 2014, the President selected the supposed Ebola czar "Ron Klain" as the White House's "Ebola Response Coordinator." Nevertheless, President Obama's decision has received wider disapproval because the so-called coordinator lacked a proper background in health care. In addition, many voters were opposing the President's decision to send the United States troops to Africa to fight the dreadful disease.

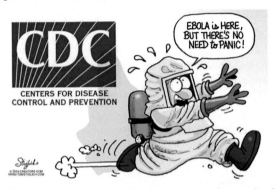

Since the United States midterm elections are coming up in October 2015, both political parties are using the events surrounding the national and international Ebola virus disease occurrence to disrepute and blame one another. Predictably, Ebola virus disease has been a burning subject in the congressional electoral race, where, for example, Senator Mark Udall of Colorado accused the Republican Party for the poor response of health care workers and emphasizing Republican-supported financial cuts of public health care agencies like the CDC, whose principal job is to forecast and respond to major health crisis.[19] On the other hand, the Republicans are accusing the Obama government for not taking necessary steps against Ebola virus disease spread. The White House has subsequently announced the formation of Special Weapons and Tactics (SWAT) teams to contain any further outbreaks if such events were to occur in the United States.[20]

REFERENCES

1. *Meet mansa musa—the richest human being in all history—newsX celebrity networth.* 2014. [cited 2014]. Available from: http://www.celebritynetworth.com/watch/e8kYOphEpmQ/meet-mansa-musa-richest-human/.
2. Elvin Mirzayer CFA F. *Ebola's economic impacts on Liberia, Sierra Leone and Guinea.* 2014. [cited 2014]. Available from: http://www.investopedia.com/articles/investing/101314/ebolas-economic-impacts-liberia-sierra-leone-and-guinea.asp.
3. Miller JW. ArcelorMittal blames Ebola for delay in Liberia mine expansion virus triggers force majeure at profitable iron ore operation. *Wall Str J* 2014. [cited 2014]. Available from: http://www.wsj.com/articles/arcelormittal-blames-ebola-for-delay-in-liberia-mine-expansion-1407522087.
4. Broderick J, Connolly S, Feldmann E, Hanley D, Kase C, Krieger D, et al. Guidelines for the management of spontaneous intracerebral hemorrhage in adults: 2007 update: a guideline from the American heart Association/American Stroke Association Stroke Council, High Blood Pressure Research Council, and the Quality of Care and Outcomes in Research Interdisciplinary Working Group. *Stroke* June 2007;**38**(6):2001–23. PubMed PMID: 17478736.

5. Team WHOER. Ebola virus disease in West Africa—the first 9 months of the epidemic and forward projections. *N Engl J Med* October 16, 2014;**371**(16):1481–95. PubMed PMID: 25244186.
6. BBC [cited 2014]. Available from: http://www.bbc.com/news/world-africa-29798555; 2014.
7. Informer T. *Ebola claims 4407 lives in West Africa: It is real! …… 'Global catastrophe' claims more victims in Texas and other parts of Europe*. 2015. [English].
8. Copley ASaA. *Understanding the economic effects of the 2014 Ebola outbreak in West Africa USA brookings*. 2014. [cited 2014]. Available from: http://www.brookings.edu/blogs/africa-in-focus/posts/2014/10/01-ebola-outbreak-west-africa-sy-copley.
9. Bank W. *World bank group Ebola response fact sheet United States of America*. 2014. [cited 2014]. Available from: http://www.worldbank.org/en/topic/health/brief/world-bank-group-ebola-fact-sheet.
10. Ananth V. *The economic impact of Ebola outbreak first published Ebola's economic impacts on Liberia Ebola's economic impacts on Liberia*. Friday, October 24, 2014. 04 46 PM IST2014 [cited 2014]. Available from: http://www.livemint.com/Specials/JHigBMhadAUdKndJjEZf5O/The-economic-impact-of-the-Ebola-outbreak.html.
11. Breitbart. *Mothers abandoning newborn in Liberia due to Ebola: online*. 2014. [cited 2014]. Available from: http://www.breitbart.com/Big-Peace/2014/11/07/Mothers-Abandoning-Newborns-in-Liberia-Due-to-Ebola-Outbreak.
12. Dailymail. *One-seven pregnant women die childbirth Ebola hit countries online*. 2014. [cited 2014]. Available from: http://www.dailymail.co.uk/health/article-2829867/One-seven-pregnant-women-die-childbirth-Ebola-hit-countries-medical-facilities-overwhelmed-say-charities.html.
13. Side effects of 1918 flu seen decades later. *Time Mag* 2009:1–2.
14. *Content time, Ebola outbreak: online*. 2014. [cited 2014]. Available from: http://content.time.com/time/health/article/0,8599,1929814,00.html.
15. Rohwerder B. *Impact and implications of the Ebola crisis*. 2014. Available from: http://www.gsdrc.org/docs/open/HDQ1177.pdf.
16. Studies IfS. *Ebola threatens social and political stability in affected countries*. 2014. Available from: http://www.issafrica.org/iss-today/ebola-threatens-social-and-political-stability-in-affected-countries.
17. Instiute AR. *Ebola in Sierra Leone: political leadership for a crisis?* 2014. Available from: http://www.africaresearchinstitute.org/blog/ebola-sierra-leone-politics/.
18. Mukpo A. *The biggest concern of the Ebola outbreak is political, not medical*. 2014. Available from: http://america.aljazeera.com/opinions/2014/8/ebola-virus-liberiasierraleonepolitics.html.
19. Jang J. *Political implications of the Ebola outbreak*. 2014. Available from: http://www.cornell-current.com/2014/10/17/political-implications-of-the-ebola-outbreak/.
20. Myrow SA. *Will Ebola affect the outcome of the election?* 2014. Available from: http://www.cnbc.com/id/102098189.

Index

Printed in the United States
By Bookmasters